ENFORCING POLLUTION CONTROL LAWS

ENFORCING POLLUTION CONTROL LAWS

Clifford S. Russell
Winston Harrington
William J. Vaughan

Resources for the Future
Washington, D.C.

© 1986 Resources for the Future

All rights reserved. No part of this publication may be reproduced by any means, either electronic or mechanical, without permission in writing from the publisher.

Printed in the United States of America

Published by Resources for the Future, Inc.
1616 P Street, N.W., Washington, D.C. 20036
Books from Resources for the Future are distributed worldwide by The Johns Hopkins University Press

Library of Congress Cataloging-in-Publication Data

Russell, Clifford S.
  Enforcing pollution control laws.

  Bibliography: p.
  Includes index.
  1. Environmental policy—United States.
  2. Environmental monitoring—United States.
  I. Harrington, Winston. II. Vaughan, William J.
  III. Resources for the Future. IV. Title.
  HC110.E5R87   1986   363.7′063′0973   85-43554
  ISBN 0-915707-25-X

The paper in this book meets the guidelines for permanence and durability of the Committee on Production Guidelines for Book Longevity of the Council on Library Resources.

Resources
FOR THE FUTURE

DIRECTORS

M. Gordon Wolman, *Chairman*, Charles E. Bishop, Anne P. Carter, William T. Creson, Henry L. Diamond, James R. Ellis, Robert W. Fri, Jerry D. Geist, John H. Gibbons, Bohdan Hawrylyshyn, Thomas J. Klutznick, Franklin A. Lindsay, Richard W. Manderbach, Laurence I. Moss, William D. Ruckelshaus, Leopoldo Solís, Carl H. Stoltenberg, Russell E. Train, Barbara S. Uchling, Robert M. White, Franklin H. Williams.

HONORARY DIRECTORS

Horace M. Albright, Hugh L. Keenleyside, Edward S. Mason, William S. Paley, John W Vanderwilt

OFFICERS

Robert W. Fri, *President*
John F. Ahearne, *Vice-President*
Edward F. Hand, *Secretary-Treasurer*.

RESOURCES FOR THE FUTURE (RFF) is an independent nonprofit organization that advances research and public education in the development, conservation, and use of natural resources and in the quality of the environment. Established in 1952 with the cooperation of the Ford Foundation, it is supported by an endowment and by grants from foundations, government agencies, and corporations. Grants are accepted on the condition that RFF is solely responsible for the conduct of its research and the dissemination of its work to the public. The organization does not perform proprietary research.

RFF research is primarily social scientific, especially economic, and is concerned with the relationship of people to the natural environment—the basic resources of land, water, and air; the products and services derived from them; and the effects of production and consumption on environmental quality and human health and well-being. Grouped into three research divisions—Energy and Materials, Quality of the Environment, and Renewable Resources—staff members pursue a wide variety of interests, including food and agricultural policy, forest economics, natural gas policy, multiple use of public lands, mineral economics, air and water pollution, energy and national security, hazardous wastes, and the economics of outer space. Resident staff members conduct most of the organization's work; a few others carry out research elsewhere under grants from RFF.

Resources for the Future takes responsibility for the selection of subjects for study and for the appointment of fellows, as well as for their freedom of inquiry. The views of RFF staff members and the interpretations and conclusions of RFF publications should not be attributed to Resources for the Future, its directors, or its officers. As an organization, RFF does not take positions on laws, policies, or events, nor does it lobby.

This book is the product of RFF's Quality of the Environment Division, Paul R. Portney, director. Clifford S. Russell is the director of Vanderbilt University's Institute for Public Policy Studies; at the time this book was written, he was the director of the Quality of the Environment Division at RFF. Winston Harrington is a fellow at RFF. William J. Vaughan, who was a staff member at RFF for fifteen years, is currently an economist with the Inter-American Development Bank. The book was edited by Nancy Lammers and designed by Elsa Williams and Martha Bari. The charts were drawn by Arts and Words. The index was prepared by Lorraine and Mark Anderson.

Contents

TABLES

FIGURES

Preface

For many years, Resources for the Future has been known as a center for research on alternatives to direct regulation in environmental management. Probably the best-known research coming out of RFF in this field has been the advocacy of effluent or emission charges by Allen Kneese. His early writings on this subject have inspired many other economists to investigate and evaluate the characteristics of such policy instruments when used in various settings.

One of the criteria called on by some researchers in their evaluations is the ease (or difficulty) of monitoring the performance of dischargers under the policy instrument and enforcing the intended behavior of the dischargers, whether it be installation of some technology, maintenance of some discharge level, or honest reporting of actual discharges. Some analysts have asserted, for example, that imposing charges on polluters presents a tougher monitoring and enforcement problem than standard methods of management—often referred to as command-and-control regulation. But others have asserted virtually the opposite: that a charge system would be "self-enforcing." Certainly there is room here for clarification.

Monitoring and enforcing pollution control laws has another and more immediate claim to attention as a policy issue in its own right. This claim arises because evidence is accumulating, principally in surveys undertaken by the U.S. General Accounting Office for the Congress, that inadequate effort and attention are being paid to these aspects of transforming our good pollution control intentions into reality. One way of stating the problem implied by this evidence is to ask what is now being done and whether it is possible to do better while remaining within the constraints imposed by limited budgets.

Moreover, the problem of monitoring and enforcement, if not quite ignored by environmental economists, has received only a tiny amount of their attention. Indeed, much of the environmental literature implicitly has assumed that dischargers will comply regardless of their self-interest. Thus, clarification of the debate over policy instruments seems to require some effort at strengthening the conceptual base for assessing monitoring programs.

Our recognition of this situation grew out of a research project that set out to study the use of economic incentives as alternatives to direct regulation, funded in 1979 by the Alfred P. Sloan Foundation. The recognition in turn led us to concentrate our attention on monitoring and enforcement. The resulting effort involved the examination of legal, technical, and statistical, as well as economic, issues. Our aims have been, first, to bring these issues together with an account of current practice in such a way as to define "the monitoring and enforcement problem" and, second, to suggest specific directions for further research.

This book has been a long time in the writing. Indeed, at times it must have seemed to supporters and colleagues that the project would never see completion. We offer our deepest gratitude, therefore, to those who stuck with us over the several years it took for the initial idea to be transformed into this book. Most important among these is the Sloan Foundation and, in particular, its vice president, Arthur Singer.

The Sloan Foundation provided RFF with the grant to study alternatives to direct regulation in environmental policy, out of which grew *Enforcing Pollution Control Laws*. Through its matching grant to RFF for environmental quality research, the Andrew W. Mellon Foundation supplied additional critical support.

We have also been encouraged by our many colleagues within RFF. Emery Castle, then RFF's president, was patient with delays and enthusiastic about specific ideas. Early on, Mark Sharefkin and Henry Peskin suggested the relevance of game theory to monitoring and enforcement problems, especially the game-theoretic models from the strategic arms literature. John Mullahy read parts of the book at different stages and offered us his helpful comments, as did Walter O. Spofford, Jr. Our heaviest debt, however, goes to Donald N. Dewees who, as a visiting Gilbert F. White Fellow, read the entire manuscript in its penultimate version and made superb suggestions for restructuring and, we hope, broadening the appeal of an inherently technical book.

Others outside of RFF influenced the shape and direction of the final manuscript. Parts of chapters 2 and 4 relied heavily upon Winston Harrington's doctoral dissertation and had been reviewed by his committee at the Department of City and Regional Planning at the University of North Carolina. Gorman Gilbert, Milton Heath, Edward Kaiser, and David Moreau, the members of the committee, are due our thanks. Chapter 6 is a shortened version of a paper presented by Vaughan and Russell at the Seventh Symposium on Statistics and the Environment in October 1982, which was published subsequently in *The American Statistician*. Chapter 7 is based on a paper delivered by Russell at the Conference on Energy and Environmental Economics held at Yxtaholm, Sweden, in August 1984. Participants at the conference made useful suggestions that are reflected in this published version.

We would also like to acknowledge the research assistance of Julia Allen, who amassed the information on monitoring technology and ran the survey reported in chapter 2 and its appendix; and Susan Bishop, who did the legal research reflected in chapter 3. This version of the manuscript reflects the helpful comments of three anonymous reviewers and of John Ahearne, vice president of RFF. We are also grateful for the editing of Nancy Lammers, whose many suggestions greatly improved the style and presentation of the manuscript. Finally, we appreciate the enthusiastic help of Betty Cawthorne, who took her usual intensely personal care and interest in the welfare of our manuscript.

February 1986   Clifford S. Russell
                                                                Winston Harrington
                                                                William J. Vaughan

1
Introduction to the Problem

Whenever laws and regulations require individuals or corporations to act in ways contrary to their self-interests, it becomes necessary to provide authority and resources to the government for monitoring and enforcement; that is, some effort must be made to observe the actions of those subject to the law. Possibilities for punishing or for at least making credible threats to punish violators of the regulations also must be available. These common-sense observations apply equally to laws intended to discourage the use or abuse of dangerous drugs, to local housing codes, to regulations written under the Occupational Safety and Health Act, and to regulations of state environmental agencies and the U.S. Environmental Protection Agency (EPA).

In any particular setting there will exist particular technical difficulties or opportunities, and even special quirks of the applicable case law, that together will define the specific monitoring and enforcement problem to be solved by the responsible agency. But, even if they are difficult or impossible to answer, the basic questions can be stated simply. How much public money and effort is it optimal to put into looking for violations and carrying through the punishment process? And what punishments fit the crime, in the sense of leading to an optimal rate of compliance now and in the future? Unfortunately, neither of these questions can be answered satisfactorily because it is not known:

- how much damage a violation causes—either directly, or indirectly when an unpunished violator encourages others to defy the law

1

- how sources will react to the prospect of uncertain punishments that will be determined by a complicated set of probabilities of detection, prosecution, and conviction
- how much detection probabilities are increased by increases in monitoring budgets.

Nonetheless, the conceptual work on crime and punishment of Becker (1968), Stigler (1970), and McKean (1980) can be helpful in keeping straight what ideally should be known and in avoiding extreme solutions, such as assuming that zero violations must be the goal or that only constant surveillance and draconian punishments can produce satisfactory results.

Two other general features of the monitoring and enforcement problem are especially important. One is the legal setting for such activities broadly defined. A second is the often-unrecognized but inevitable problem of uncertainty that is inherent in measurement or observation and the resulting probabilities of missing actual violations and of making false accusations. These errors cannot be entirely avoided, but they can be reduced by expending more effort. Thus, the simple question about monitoring effort really must be a sophisticated one involving probabilities of failure to find violations and of the identification of false violations.

This book is about monitoring and enforcement in the context of pollution control laws, and its specific methods and conclusions reflect that context. But many of its conceptual underpinnings, in particular the normative models of monitoring and enforcement policy ultimately developed, apply to other areas of public policy as well. Thus, the reader with an interest in keeping contaminated food products off supermarket shelves or in encouraging safe and healthy work places will find lessons in what follows, even if the examples used relate to pollution discharges.

Monitoring and Enforcement in the Context of Pollution Control Policy

The point of pollution control is to make the everyday, outdoor environment a more healthful and pleasant one. This ultimate goal has been obscured in the United States, however, where, in reaction to earlier failures, pollution control legislation and regulation have concentrated minds and money on installing technology. The short-term goal has been to reduce pollution by placing upper limits on allowable discharges and forcing the installation of equipment that is capable of attaining those discharge levels. Neglected in the push for visible achievement have been three especially important long-term issues. The first of these is the net benefit to society—whether the existing set of pollution reduction goals represents an even roughly efficient way to proceed. The second is the future of technology—whether the current approach encourages or discourages decentralized research and development that will lead to a better environment at the same cost tomorrow. And the third is continuing monitoring and enforcement—whether the existing system can provide incentives for continuing compliance by dischargers with whatever set of standards (or pricing schemes) are in force.

The first of these issues has been and continues to be a subject of great interest. Environmental economists especially, but not exclusively, have been drawn to it and have written about alternatives to existing approaches (See, for example, Rose-Ackerman, 1973; Freeman, 1982; and Tietenberg, 1980). Some of this work also relates to the second issue, for it has been shown that different ways of ordering or inducing individual dischargers to do what is required to achieve desired ambient quality goals produce different incentives for efforts to improve pollution control technology (Bohm and Russell, 1985; Magat, 1978; Wenders, 1975). The third issue, in contrast, has been largely, though by no means entirely, ignored by economists and by other policy analysts.

Is Monitoring and Enforcement Really an Issue?

One common thread running through the literatures on efficiency and incentives for technical change is the assumption, implicit or explicit, that polluters, in fact, will comply with the discharge standards they are issued or will pay an accurately drawn bill if charging for emissions is the alternative under discussion. What is almost always missing is examination of this important, one might say vital, assumption. Given the policy fixation outlined above, perhaps this is not very remarkable. But it does seem to fly in the face of another basic assumption of the efficiency literature: that dischargers are motivated by self-interest in responding to whatever orders or charges are in place.

If the world is replete with examples of the pursuit of self-interest leading to the violation of laws, how could so many sound and sober thinkers have ignored this possibility in the pollution control field? A key to the answer lies in another assumption—almost always implicit in the writings in question—that the responsible agency knows just what each individual source is discharging at all times. Thus, the assumption of perfect (and, incidentally, costless) monitoring has supported the assumption of perfect compliance.

A more practical and much smaller body of literature has devoted attention to alternatives to perfect agency monitoring as the motivation for compliance by discharge sources. One theory asserts that sources will try to comply even in the absence of fines for violations. This view, which appears to be common among those actually engaged in enforcement at the state and federal level, essentially rests on the argument that dischargers have many reasons to obey the discharge limits imposed on them or to correct any discovered violations. These reasons include the desire to create a positive public image and not to provoke a broader bureaucratic attack, which might include tax audits or blacklisting on government contractor lists, by being discovered in flagrant violation of one set of regulations. This view is examined more formally in chapter 4.[1]

Another theory is based on an analogy with the U.S. self-reporting system of income taxation. In this view, an acceptably accurate knowledge of actual emissions can be achieved by requiring all sources over some designated threshold to measure their own discharges and report them to the responsible agency. As with the Internal Revenue Service (IRS), the agency need only conduct audits, whether random or directed by the lessons of experience and the characteristics of past violators, to encourage accurate reporting. Violations discovered in the audits would be penalized in administrative, civil, or criminal proceedings depending on severity and evidence of intent.

Two observations can be made immediately about this line of thought. The first is that its major feature, self-monitoring and -reporting, is already a central part of the current system. As will be seen in chapter 2, almost all state environmental agencies rely heavily on self-reporting by polluters for knowledge of what is being discharged. Requirements for such efforts are written into individual water pollution permits under the National Pollution Discharge Elimination System (NPDES) and into the New Source Performance Standards (NSPS) governing new sources of air pollution (See Wasserman, 1984, tables IIIa-1 and IIIa-4).

The second observation, however, is that the IRS has two advantages in administering the U.S. system of self-reporting income taxation that an environmental agency inevitably and unavoidably will lack:

- The IRS has access to an extensive paper record covering most

---

[1] This approach has been referred to in some places as "voluntary compliance," which implies a situation slightly at variance with the one described; that is, compliance is not entirely voluntary, since there are sanctions held in reserve. Furthermore, "voluntary compliance" will be reserved in this book to refer to a system of enforcement in which sources discovered in violation are not punished immediately but rather are allowed a chance to return to compliance without penalty. This use, too, is slightly misleading, but it does have the advantage of being common.

if not all income for nearly all individuals. Its audits consist largely of checking that record against self-reported income and deductions. Where the IRS's own records (W-2s, 1099s, and so on) are inadequate, it can require proof from the taxpayer that claimed transactions really took place and can check bank account transactions for evidence of unreported income.

- For the largest part of taxable income, it is to the advantage of taxpaying firms or individuals to make sure that the IRS has a complete record for other firms and individuals, because one taxpayer's income is another taxpayer's deduction.

In contrast, for pollution control agencies, there is no complete, independent record of what self-monitoring dischargers are emitting. Once discharges have gone up the stack or out the pipe they have vanished from an enforcement point of view and have left no record in the world. In this sense, they are fugitive events. Even if atmospheric discharge concentrations can be measured by remote monitoring equipment (Williamson, 1981) or source discharges can be inferred from measured ambient quality levels and discharge compositions (see Gordon, 1980; Courtney, Frank, and Powell, 1981), it still is not possible for the agency to monitor on its own schedule. Rather, it must act as discharges occur, or it loses the chance of acting at all. Furthermore, it is generally not in the interest of any particular firm or individual to provide independent evidence of discharges by another source routinely. While someone may be motivated by bounties or simple outrage to report obvious offenses, individuals cannot be counted on to do more than spot the tip of the iceberg. They lack both access to premises for in-stack monitoring and technology that would allow them to measure anything but the very crudest indicators, such as plume opacity for air pollution sources. (As discussed in chapter 8, however, the self-monitoring reports can be used by interested individuals and groups in enforcement actions.)

Some Evidence

The impatient reader may be unimpressed with these arguments and counterarguments. A priori debate about incentives and possibilities is one thing, but where is the evidence that long-term monitoring and compliance *ought* to be an issue? Maybe an ad hoc combination of moral suasion and self-monitoring really is good enough.

This is an understandable, if slightly unfair, position. It is almost impossible to know how common compliance is in the absence of serious monitoring efforts, for the behavior of the polluting companies would change if they knew they were being monitored. Only by contriving some elaborate and highly artificial independent and confidential measurement could one observe what sources do in the absence of fear of discovery by the responsible environmental protection agency. A priori argument based on the profit-maximizing version of self-interest with limits on the influence of altruism or public relations concerns must therefore remain a major justification for the view that monitoring and enforcement is a problem.

But this is not to say there is *no* evidence on continuing compliance.[2] The studies of both Harrington (1981) and McInnes and Anderson (1981) suggest that a significant fraction of point sources of air pollution are out of compliance for substantial parts of every year. Even more impressively, a 1983 study by the U.S. General Accounting Office (GAO) of wastewater dischargers concludes that failure to comply with permit discharge limits is widespread; the full extent of noncompliance may not be known; and current enforcement practices do not encourage prompt correction of noncompliance after its discovery.

More specifically, the GAO study team reviewed discharge monitoring reports from 531 major wastewater dischargers (roughly half industrial and half municipal) to discover the extent of self-reported violations of permit terms during an eighteen-month period ending 31 March 1982. Furthermore, the team reviewed the completeness of the self-monitoring reporting for each of the sources during the same eighteen-month period. For one subsample of dischargers, GAO explored how long noncompliance continued before enforcement action was taken.[3]

Overall, 31 percent of the dischargers examined were found to have been in significant noncompliance during the period. ("Significant noncompliance" was taken to mean exceeding permitted concentration or quantity limits by 50 percent or more for at least one quality parameter in at least four consecutive months.) The rates were 28 percent for municipal and 21 percent for industrial sources. That these rates might be underestimates of the overall noncompliance is suggested by the fact that 8 percent of the sources failed to submit one or more of the reports required of them during the period, while 37 percent submitted one or

---

[2] As will be emphasized later in this chapter, "compliance" within the current system can have two meanings. The more common might be restated as having the ability to meet discharge standards as embodied in permits. "Compliance schedules," for example, refer to time tables for installation and approval of equipment that will allow permit terms, written from guideline documents, to be met. This is called "initial" compliance in what follows. The actual meeting of these terms on a day-to-day, week-to-week, or month-to-month basis is the meaning commonly used in this book. This situation will be referred to as "continuing" compliance.

[3] Questions about permit backlogs and expired-but-not-reissued permits were also explored.

more incomplete reports. The GAO report also observes that efforts to check up on self-monitoring are being reduced by the EPA and the states, with data on sampling inspection cut back. Finally, in commenting on the reaction produced by reported significant noncompliance with permit terms, GAO says: "In some cases, formal enforcement action was not taken for years after noncompliance began. In other cases, noncompliance had continued for years even after EPA or the state took enforcement action."

A Source of Confusion:
Initial Versus Continuing Compliance

A student of pollution control policy might be tempted to object to the above recitation of evidence and to cite apparently contradictory EPA evidence that shows very high rates of compliance indeed. For example, according to data from EPA's Compliance Data System, more than 90 percent of major stationary sources of air pollution had achieved compliance with state regulatory requirements as early as the middle of 1977.[4]

---

[4] That left nearly 1,500 sources that had not achieved compliance or were not even on an approved compliance schedule. Moreover, those not yet in compliance tended to be very large sources and were concentrated in the electric power and steel industries (U.S. GAO, 1979). For a number of reasons, these sources presented EPA with a problem. Owners of large facilities often had both the incentives and the means to fight hardest against regulation. They also frequently presented the state agencies with a political problem—especially those sources located in small towns where they provided a large share of local employment. In this context, the 1977 Amendments to the Clean Air Act gave EPA the authority to seek civil penalties (up to $25,000 a day) and administrative noncompliance penalties. Although the courts would remain reluctant to take vigorous enforcement action, the agency at least had a credible threat to make against those choosing not to comply (See Melnick, 1983).

By mid-1981 EPA civil actions had been initiated against 395 facilities, and state civil actions against 78. In addition, about a thousand sources were subject to state or EPA administrative action. Fully half the 1,500 noncomplying sources had achieved initial compliance, while another quarter were on a schedule for achieving compliance within a short while (memo issued May 11, 1981, from Richard Wilson, acting assistant administrator for enforcement, to regional administrators). Most of these cases, therefore, were concluded successfully.

This improvement in compliance during the late seventies was achieved with virtually no actual use of legal penalties, although large penalties were threatened by EPA. Between 1977 and 1981, $27 million in penalties of all types was collected by EPA for violations of the stationary source regulations (Crandall, 1983). State agency collections were even less. These penalties are insignificant, especially when compared with estimated annual investment costs of $3.9 billion for stationary source abatement (U.S. CEQ, 1979).

A critical look at compliance data gathered by EPA is provided by Wasserman (1984). From her data it appears that in 1983, 92.9 percent of existing sources of air pollution with potential emissions of more than 100 tons each year were considered either "in compliance" or "on schedule" toward achieving compliance. The rates for water pollution sources (which must be read from a small, rough graph) were approximately 96 percent for major private industrial sources and 81 percent for major publicly owned sources. ("Major" sources are the largest 15 percent in the national inventory.) As Wasserman explains, however, these data do not reflect compliance with permit terms on a continuing basis.

Thus, there is a problem of definition. Because of its required interest in encouraging the installation of technologies designed to produce desired discharge reductions, EPA historically has defined compliance in those terms; that is, sources are "in compliance" when they have installed and demonstrated (or are installing or have agreed to install) the appropriate equipment. For the purposes of this book—and, it would seem, for clarity in future debate—this type of compliance will be referred to as *initial compliance*. It is simply a successful demonstration that a particular plant or part of a plant is capable of meeting a required limit on discharges. *Continuing compliance*, in contrast, is the ongoing meeting of a discharge limit over days, weeks, and years of routine operation.

The problems an agency has in enforcing initial compliance are considerably different from those for continuous compliance, and a listing of three major differences may illuminate the difficulties of enforcing continuous compliance:

- Achievement of initial compliance usually has no effect in itself on environmental quality. Initial compliance can be satisfied by obtaining and operating correctly in one test the appropriate equipment, but environmental quality will not be affected unless the equipment operates continuously. One exception to this rule occurs when the source complies with the regulation by eliminating the regulated pollutant altogether.
- Achieving initial compliance generally involves assuming a fixed cost, often through the purchase of some major piece of equipment or the modification of a major process. A source that successfully postpones initial compliance postpones this expenditure. Once initial compliance has been achieved, however, avoidance of continuing compliance saves at most a portion of the variable operating cost.
- Once initial compliance has been achieved, that is that; imple-

mentation is a one-shot operation. Monitoring simply requires a single test of the capability of the installed equipment (See, for example, Downing and Watson, 1973, 1974, 1975). Surveillance and monitoring for continuing compliance is a continuing problem.

These differences raise and then answer the question of why enforcement authorities bother with initial compliance at all. After all, if environmental quality is not necessarily improved, why not go directly to continuous compliance? The reasons are that initial compliance is relatively easy to monitor; that it is a useful precursor to, if not a necessary condition for, continuous compliance; and that its achievement greatly reduces the incentives for subsequent noncompliance. Thus, even if initial compliance were not a formal part of implementation programs (through issuing permits), agencies almost certainly would develop some similar concept and apply it informally.

Another question suggested by these observations then might be, why does this book not bother more with initial compliance, given its importance for subsequent continuing compliance? The answer, briefly, is that initial compliance with the major U.S. pollution control regulatory programs has been nearly achieved, as described earlier; while, on the contrary, the task of ensuring continuing compliance stretches endlessly before us.

Fragmentary empirical evidence seems to support the contention that insufficient effort is now being applied to monitoring and enforcing continuing compliance with discharge permits. Furthermore, powerful a priori arguments can be made against the contention that voluntary compliance and self-monitoring in themselves can carry the weight of a very expensive pollution control system into the coming decades. While analytical effort is being expended to improve on the current regulatory implementation system, corresponding effort should be devoted to the question of what an improved monitoring and enforcement system would look like. This book is intended as a step in that direction—far from the last word, but at least catholic in its definition of the problem.

Complications

Catholicity is especially important in coming to grips with the issues of monitoring and enforcement because of the complex interplay between the legal, statistical, and technological aspects of the problem. The economist or policy analyst runs a great risk by trying to assume away any of these dimensions. But including them all, and especially allowing for the statistical complications inherent in actual situations, creates serious difficulties for modeling and systematic thinking.

One might suppose that to monitor pollution-control performance, inspectors need merely walk into an industrial plant or other point source, insert some perfectly accurate meter into the wastewater or smoke stack emission stream, and read a constant rate-of-discharge result that could be expressed equally well in any time unit, from per second to per year. In reality, the inspector's first challenge is to get into the plant at all. As discussed in chapter 3 constitutional questions about the right of access for monitoring purposes have not yet been answered satisfactorily, but current practice appears to be to announce visits in advance, with all that implies for cheating possibilities.

Once inside, the problem of unwieldy and expensive sampling equipment comes to the fore. (This equipment was designed for *initial* compliance testing, when preannounced visits make sense so that long set-up times are not a problem.) The complicated equipment requires that the process of measurement will (a) take significant amounts of set-up time over and above time spent actually sampling (sometimes as much as several days), and (b) reflect discharge rates during short periods of actual sampling (that is, several hours at most). Thus, the testing is costly and the information gained is small relative to what would be contained in a complete trace of the source's discharges over a month or year.[5]

The measurement difficulties are important because even when the source is trying to comply with its permit terms it will have fluctuating discharges. These fluctuations may have both periodic elements due to production patterns, boiler soot-blowing, or other routine causes as well as random components ultimately traceable to human or machine failure, fluctuations in ambient conditions such as temperature, or random startup and shutdown decisions. Thus, the stream being measured is not constant, and measurements at one time can be applied only to broader compliance questions through statistical inference with associated probabilities of errors of two types—that a violation will be found where none exists or that true violations will be missed. To further complicate matters, the measurement instruments have their own errors that must be accounted for in the inference procedure. The perfect instrument does not exist.

Finally, as has been pointed out above, it is almost certainly misleading

[5] This discussion ignores the existence of in-place continuous monitoring equipment. The procedures described might be necessary to check up on such equipment in any case.

to assume that sources in general are trying to comply with the standard or to establish an accurate basis for an emission charge bill. Cheating cannot be ruled out.

Logically subsequent to but intimately connected with the monitoring problem is enforcement. If a violation is discovered (with certain probability), what should the agency do? Should it simply inform the source and request correction? Must it invoke a fine? Should the structure of fines reflect the size of the violation? Should the source's history of violations count either in determining the fine or in deciding about monitoring? These issues must also be considered in assessing current policy and suggesting alternatives for the future.

The Plan of the Book

This book grew out of work on using economic incentive as an alternative to strictly regulatory approaches in pollution control. In that research it became clear that whenever ease or difficulty of monitoring and enforcement was mentioned as a basis on which to compare two alternatives, the conclusions tended to be ad hoc and their support weak or nonexistent. It seemed that a substantial void existed in the environmental policy literature; this book represents a first step in filling it.

In writing this volume, the authors developed a substantial sympathy for those who had confined themselves to speculation or casual assertion about monitoring and enforcement. The subject is difficult, and to do even rough justice to it requires giving some attention to the legal complications and to the implications of the characteristics of monitoring instrumentation for the ability of the environmental management agency to come to grips with stochasticity and intentional violation. Furthermore, it seemed unfair and misleading to write about what ideally ought to be done without taking some notice of what actually is being done.

All these considerations led inexorably to this volume. But before its structure is set out, it will be well to be clear about what is *not* attempted in it. First, this book does not cover the interesting and difficult problem of designing ambient environmental monitoring systems. (On this subject, the interested reader is referred to a 1977 study on environmental monitoring conducted by the National Academy of Sciences and, more recently, to issues of the new journal *Environmental Monitoring and Assessment,* for example, Pickett and Whiting, 1981.) Second, there is no claim here to completeness. Rather, the goal of the book is to provide a broad perspective for the analysis of the monitoring and enforcement problem.

Chapter 2 concentrates on two major questions: What type and how much monitoring is being done? What enforcement policies are pursued to follow up on the results of the monitoring? To answer these questions, the state (and to some extent local) environmental agencies are examined.

Chapter 2 suggests rather strongly that the responsible agencies do very little actual measuring of discharges, but that they do, through permits, require polluters to monitor themselves and to report violations. Furthermore, chapter 2 shows that actual fines are not usually imposed to penalize violators. A significant fraction of state enforcement efforts rely on "voluntary compliance," in which detected violations trigger warnings that offer a chance to remedy the situation without penalty.

Chapter 3, which presents further background on the legal and technological aspects of monitoring and enforcement, addresses the following questions:

- How has the law—in particular how have the courts—dealt with the inevitable statistical errors referred to earlier?
- How does the law stand in the matter of gaining entry to a polluter's premises for the purpose of monitoring?
- How, if at all, can the development of remote (off-premises) monitoring equipment help the agencies with their responsibilities?
- What, roughly, are the costs and capabilities of available monitoring technologies?

Chapter 4 turns to the economics of monitoring and enforcement. First, it reviews the economic and policy literatures within a framework stressing the importance of assumptions about the presence or absence of random elements in the discharge pattern (or random measurement errors, or both) and whether sources can choose to violate their permit terms or only do so by inadvertence. The second part of chapter 4 provides a very simple model designed to mimic "voluntary compliance"; that is, sources are assumed to be trying to comply but are subject to random equipment failures leading to violations, whether discovered or not. But discovery of a violation, if one is occurring, is assumed certain *if* the agency conducts a monitoring visit at that time. No statistical problems exist. Furthermore, the source is assumed to correct the situation upon its discovery by the agency, without the imposition of any penalty. It will be seen that, under plausible assumptions about the parameters involved, the extent of noncompliance in the model is not wildly different from that suggested by the studies cited in this chapter.

The subsequent three chapters introduce the complications of random errors, and thus agency uncertainty, and of willful noncompliance. Chapter 5 outlines some necessary, if perhaps painfully complicated, statistical background. In chapter 6, that background is used to develop a model for designing monitoring schemes based on statistical quality control methods—methods that are most useful when willful noncompliance is still ruled out and that require knowledge of the benefits of controlling particular sources.

Quite a different tack is taken in chapter 7. There it is allowed that sources may violate intentionally, and a rather different methodology—single- and multiple-play games—is used to model this situation. This normative model can be used to guide the agency in how best to use scarce resources for monitoring and in how to relate its monitoring efforts to its enforcement choices, in particular the size of the fines it seeks for violations. It turns out that by retreating from the requirement that a monitoring and enforcement scheme optimally balance society's costs with the benefits produced, it is possible to fashion a workable policy design. To anticipate that result in sketchy form, it will be shown that by grouping sources for monitoring purposes on the basis of past detected behavior, it is possible to reduce noncompliance to very low levels while staying within tight monitoring budget limitations.

The final chapter summarizes the lessons to be found in the book—matters both of current performance and future possibilities. The chapter also stresses, however, the extent to which the conclusions are tentative. The sources of this tentativeness are examined for their implications for further research.

References

Becker, Gary S. 1968. "Crime and Punishment: An Economic Analysis," *Journal of Political Economy* vol. 76 (March/April) pp. 169–217.

Bohm, Peter, and Clifford S. Russell. 1985. "Comparative Analysis of Alternative Policy Instruments," in Allen V. Kneese, ed., *Handbook of Environmental and Resource Economics* (New York, North Holland).

Courtney, F. E., C. W. Frank, and J. M. Powell. 1981. "Integration of Modelling, Monitoring, and Laboratory Observation to Determine Reasons for Air Quality Violations," *Environmental Monitoring and Assessment* vol. 1, no. 2, pp. 107–118.

Crandall, Robert. 1983. *Controlling Industrial Pollution* (Washington, D.C., Brookings Institution).

Downing, Paul B., and William D. Watson, Jr. 1973. *Enforcement Economics in Air Pollution Control,* EPA 600/5-73-014 (December) (Washington, D.C., U.S. Environmental Protection Agency).

———. 1974. "The Economics of Enforcing Air Pollution Controls," *Journal of Environmental Economics and Management* vol. 1, pp. 219–236.

———. 1975. "Cost-Effective Enforcement of Environmental Standards," *Journal of the Air Pollution Control Association* vol. 25, no. 7 (July) pp. 705–710.

Freeman, A. Myrick III. 1982. *Air and Water Pollution Control: A Benefit-Cost Assessment* (New York, John Wiley & Sons).

Gordon, Glen E. 1980. "Receptor Models," *Environmental Science and Technology* vol. 14, no. 1, pp. 792–800.

Harrington, Winston. 1981. *The Regulatory Approach to Air Quality Management* (Washington, D.C., Resources for the Future).

Magat, Wesley A. 1978. "Pollution Control and Technological Advance: A Dynamic Model of the Firm," *Journal of Environmental Economics and Management* vol. 5, pp. 1–25.

McInnes, Robert G., and Peter H. Anderson. 1981. *Characterization of Air Pollution Control Equipment Operation and Maintenance Problems* (Washington, D.C., U.S. Environmental Protection Agency).

McKean, Roland N. 1980. "Enforcement Costs in Environmental and Safety Regulation," *Policy Analysis* vol. 6, no. 3, pp. 269–289.

Melnick, R. Shep. 1983. *Regulation and the Courts: The Case of the Clean Air Act* (Washington, D.C., Brookings Institution).

National Academy of Sciences. 1977. *Environmental Monitoring, A Report to the U.S. Environmental Protection Agency from the Study Group on Environmental Monitoring* (Washington, D.C., National Academy of Sciences, National Research Council).

Pickett, E. E., and R. G. Whiting. 1981. "The Design of Cost-Effective Air Quality Monitoring Networks," *Environmental Monitoring and Assessment* vol. 1, no. 1, pp. 59–74.

Rose-Ackerman, Susan. 1973. "Effluent Charges: A Critique," *Canadian Journal of Economics* vol. 6, pp. 512–528.

Stigler, George J. 1970. "The Optimum Enforcement of Laws," *Journal of Political Economy* vol. 78, pp. 526–536.

Tietenberg, T. H. 1980. "Transferable Discharge Permits and the Control of Stationary Source Air Pollution: A Survey and Synthesis," *Land Economics* vol. 56, no. 4 (November) pp. 391–416.

U.S. Council on Environmental Quality. 1979. *Environmental Quality—1978* (Washington, D.C., U.S. Government Printing Office).

U.S. Environmental Protection Agency. 1978. *Civil Penalty Policy* (Washington, D.C., Office of Enforcement, April 11).

U.S. General Accounting Office. 1979. *Improvements Needed in Controlling Major Air Pollution Sources* (Washington, D.C., U.S. GAO) CED-78-15.

———. 1983. *Wastewater Dischargers Are Not Complying with EPA Pollution Control Permits* (Washington, D.C., U.S. GAO) GAO/RCED-84-53 (December).

Wasserman, Cheryl. 1984. "Improving the Efficiency and Effectiveness of Compliance Monitoring and Enforcement of Environmental Policies. United States: A National Review." Draft (Organization for Economic Cooperation and Development, Environmental Directorate).

Wenders, John T. 1975. "Methods of Pollution Control and the Rate of Change in Pollution Abatement Technology," *Water Resources Research* vol. 11, no. 3 (June) pp. 393–396.

Williamson, M. R. 1981. "SO_2 and NO_2 Mass Emission Surveys: An Application of Remote Sensing," in Air Pollution Control Association, *Continuous Emission Monitoring: Design, Operation, and Experience* (Pittsburgh, Pennsylvania, APCA).

2
Current Efforts to Induce Continuous Compliance

The major activities of environmental quality management agencies intended to promote continuous compliance fall into two categories: monitoring and enforcement. Monitoring is used to determine whether pollution sources are in compliance with applicable regulations. Enforcement is intended to deter sources from violating the regulations and to force violators to return to compliance.

Unavoidable variation in discharge levels and inevitable imperfection of measurement equipment introduce ambiguity into the very notion of compliance and noncompliance. Later chapters will explore these difficulties, but for now the discussion shall proceed as if the cost of monitoring and enforcement activities were the only problem facing the agencies.

The first section of the chapter is concerned with a more careful definition of the "applicable regulations." Since these are almost invariably expressed as the terms of individual discharge permits, one instructive exercise will be to see what a permit actually looks like, and key excerpts from one permit are presented in a table. (The full text of a permit is contained in appendix 2-A.) Then, the results of a survey of state monitoring activity undertaken by Resources for the Future (RFF) in 1982, which contain data on the kinds of requirements imposed by permits, are outlined.

The next section briefly describes the types of monitoring and enforcement options that are available to responsible agencies. Then the chapter turns to actual monitoring and enforcement practices, looking at the results from the RFF survey, at results from studies of specific states reported by others, and at more detailed information from a single state, New Mexico, studied for the U.S. Council on Environmental Quality by one of the authors of this book. (The complete RFF survey results are summarized in appendix 2-B.)

What Is Being Monitored and Enforced?

While agency and public concern is ultimately with the environmental quality citizens encounter in their daily lives, environmental monitoring and enforcement efforts normally concentrate on the behavior of individual pollution sources—what they discharge and how this compares with requirements imposed on them. Within this general framework, there is room for choice of specific regulations or standards that can be applied.

Table 2-1 summarizes the major types of standards available to legislatures and environmental management agencies. Briefly listed are the characteristics of each standard: what must be measured, at how many points measurements must be made, and how directly the standard relates to ambient quality. Other considerations are also suggested.

Some choices allow the source flexibility and require more effort from the agency. For example, a standard written to limit pollutant discharges per unit of plant input or output automatically allows the plant to expand discharges as production increases. It also obliges the agency to measure simultaneously both discharge and input or output to check for compliance. In contrast, a limit on pollutant per unit of time means that discharge per unit of input or output must be reduced if production is to be increased; the agency need not measure either input or output. Furthermore, standards differ in the directness with which they control ambient quality effects of the source. When a single source is present in a region, direct control is possible through specification of the allowable ambient effects produced by the source's discharge. This strategy will work best when a single isolated source is involved. Other choices, such as placing a limit on the concentration of pollutants in a waste stream, offer only the most tenuous control; total discharges are not regulated, making it possible to dilute waste streams to achieve concentration standards. Yet, in this instance, monitoring is relatively easy because it does not require any measurement of waste stream volume.

Within this range, which options do the agencies actually choose? The RFF survey of state agency monitoring and enforcement practices asked agencies to indicate what types of standards were included in their per-

Table 2-1. Characteristics of Pollution Control Standards

Characteristics	Types of standard					
	Pollutant concentration	Pollutant weight per time period	Pollutant weight per unit production	Fuel or other input quality	Percentage removal of pollutant	Ambient standards around source
Required to measure	Concentration only	Concentration plus volume of effluent/ emission over same time	Concentration plus volume plus output over same time	Concentration of impurities (e.g.: sulfur in fuel)	Concentration plus volume before and after removal	Ambient concentrations of pollutant or secondary product
Number of points of measurement	Minimum of one	Minimum of one	Minimum of one for discharge, one for output	Minimum of one	Minimum of two	Minimum of one; likely to be several
Relation to impact on environmental quality	Very tenuous. Total amount of discharge not controlled	Quite direct. Total amount controlled, though impact varies with external conditions such as wind speed, or water temperature.	Indirect in principle, but quite direct in short run when maximum production levels fixed.	Fairly direct in short run when maximum capacity and type of equipment fixed.	Fairly direct in short run when maximum capacity and hence raw load fixed.	Very direct
Other considerations	Opacity is a special case of concentration limit.	Monitoring for an emission/effluent charge system involves this same approach and problems.	Does not limit total emissions in long run.	Applicability depends on extent of source's choice about operating conditions and conversion rates.	Provides no incentive for reductions in raw loads.	Very hard to enforce where more than one source in small area.

Table 2-2.   Agencies Using Particular Types of Standards in Permits

	Air	Water
Total agencies reporting	30	44
Proportions using limits on:		
Concentration	0.97	1.00
Mass/unit input[a]	0.97	0.36
Mass/unit output	0.70	0.50
Mass/unit time		
per minute	0.10	0.04
per hour[a]	0.70	0.09
per day[a]	0.33	0.59
per week	0.07	0.14
per month[a]	0.07	0.27
per year[a]	0.33	0.07

Source: RFF survey.

[a]Difference in proportions between air and water significant at 5 percent level or better.

mits (but not how frequently each type might appear across all the permits issued in a state). The agencies were not asked to match type of permit terms with type of pollutants being controlled. The goal of the survey was to identify the range of monitoring problems faced by the states because of the permit terms in use, not to understand exactly how that range was created nor to estimate the full size of the monitoring chore. Thus, the figures in table 2-2 have their limitations; the proportions shown are of reporting agencies that use the particular standard in any permits. They are not the proportions of total measurements (permits times residuals) required to be of a particular type.

Notice in table 2-2 that standards on pollutant concentrations are used in permits for both media in virtually every reporting state. Only one agency reported that it used no permits in which concentrations are specified. This suggests that ease of monitoring is an important consideration when choices of standards are made.

When standards on permitted mass of emissions are examined, no such unanimity is found; in fact, interesting contrasts between the two appear. For example, both mass-per-unit-of-input and mass-per-unit-of-output standards are used very widely in monitoring air pollution sources, much less widely for water pollution.[1] A clear majority of states uses one of the available standards on mass of emissions per unit of time. Some of these are adopted quite commonly (air emissions per hour, per day, and per year as well as water emissions per day and per month). Others, such as the emissions-per-minute standard, are generally avoided.

[1] Actually, only the difference between the proportions for mass-per-unit input is statistically significant at these sample sizes, if it is required that the chance of accepting a spurious result be kept to 5 percent.

In this connection it is worth noting the interaction between monitoring instrument characteristics, standard definition, and treatment process design. For example, in most instances of air pollution, there is no storage between generation and discharge of pollutants. Thus, an instantaneous sample taken from a smokestack reflects no averaging. But the equipment and methods approved by the Environmental Protection Agency (EPA) (described more fully in the next chapter) are designed to draw a continuous sample from the stack over a period of several hours. Thus, a monitoring event might involve observation of hourly emissions for several different hours—a sample from the universe of hours to which the common hourly emission limit applies. In contrast, water pollution control equipment commonly does make use of storage and mixing so that an instantaneous sample from downstream of the storage reservoir reflects an average of instantaneous raw emission rates; that is, a slug of pollutant entering the treatment system after an accident would be discharged over a long period and would raise actual hourly discharges by much less than hourly input changed. As a rule, water pollution monitoring involves combining such instantaneous samples taken at different moments into a composite sample intended to reflect the average conditions over some period. Notice that if, for example, daily emission is constrained, no single observation actually involves measurement of total discharge over a day. Rather a sample from the set of moments making up the day are used and the implied daily emission projected from them. This contrasts with the air pollution situation.

A specific example of a pollution permit will help to make the above arguments and survey results more concrete. Table 2-3 outlines the heart of a North Dakota air pollution control permit—the emission standard set. The terms of this permit do vary as expected. Though no concentration figures are specified, standards are specified both in terms of pounds per hour and pounds per million British thermal units (Btu) of heat input.[2]

What Methods of Monitoring and Enforcement Are Commonly Used?

The state agencies responsible for checking up on compliance of pollution sources with the terms of their discharge permits have available a range of techniques for gathering information. These vary in the kinds

---

[2] See also Rossnagel (1978) on emission limits in the laws and regulations of fifty-three states and cities. And see Frye and Ayres (1981) on air pollution permits in areas subject to prevention of significant deterioration regulations.

Table 2-3.   Emission Standards from a North Dakota Air Pollution Control Permit

Source	Maximum permitted emission rates (lbs./hr.)		
	TSP	SO$_2$	NO$_x$
I. Coal handling			
a. Primary crusher station	9.4		
b. Secondary crusher station	13.4		
c. Belt conveyor transfer and sampling station 1	10.6		
d. Fines screening and crusher station	11.7		
e. Belt conveyor transfer section 2	2.7		
f. Fines storage silo and belt conveyor transfer station	1.5		
g. Gasifier building	28.3		
h. Belt conveyor and sampling station 3	13.4		
i. Live storage reclaim tunnel ventilation	2.3		
II. Refuse incinerator	1.98	Negligible	Negligible
III. Main stack[a]			
a. Normal operations	170	2,610	1,100
b. 2-hour shift catalyst regeneration	170	5,360	1,100
c. 8-hour shift catalyst regeneration	85	4,060	550
d. 7-hour startup incinerator operation	85	1,305	550
IV. Shift catalyst regeneration unit	Negligible	2,750	Negligible
V. Low-pressure flare	Negligible	218[b]	Negligible
VI. Startup incincerator	Negligible	255[c]	Negligible
VII. Steam boilers (lb/MMBTU heat input)	0.1	0.8	0.557
VIII. Superheater furnaces (lb/MMBTU heat input)	0.1	0.8	0.24
IX. Stretford off-gas stream	Negligible	12,007	Negligible
X. Main flare A	Negligible	90,408	Negligible
XI. Main flare B	Negligible	90,408	Negligible

Source: North Dakota State Department of Health.

[a]Includes total emissions from steam boilers, superheaters and the shift catalyst regeneration unit.

[b]For nonemergency conditions; for upset conditions not to exceed fifteen minutes, the maximum rate is 582 lbs./hr.

[c]During hot operation; during cold operation, the maximum rate is 145 lbs./hr.; during startup following scheduled maintenance or equipment failure, the maximum rate is 1,152 lbs./hr.

of information they produce and, in particular, in whether that information definitely shows or is merely indicative of a permit violation. The methods also vary in the extent to which the source must be asked to cooperate or at least not to interfere. Finally, the methods have different costs; and, not surprisingly, the more definitive the information, the higher the costs. In this section, a very brief catalog of monitoring methods is presented as background for the rest of the chapter and as an introduction to the subject of monitoring technology. (The latter will be returned to in chapter 3.) The second part of the section offers similarly brief comments on enforcement methods.

Monitoring Alternatives

The simplest, cheapest, and most common monitoring method is the inspection, in which a trained observer attempts to infer compliance status from indirect evidence, such as equipment-operating logs, plant and equipment appearance and state of repair, and the visual quality of the current discharges. Strictly speaking, an inspection cannot determine compliance status except where compliance is defined specifically in terms of visual findings, as it is for opacity of stack gas discharge, or where compliance is defined in terms of equipment-operating parameters such as pressure drop or stack gas temperature. To be sure, the inference of a violation sometimes can be made with a high degree of confidence. In New Mexico, for example, the air pollution emission regulations for asphalt and nonmetallic mineral processors are set sufficiently low that a visible emission probably represents a violation. For pollutants such as dissolved solids or pH (in wastewater) or SO_2 or NO_x (in air), examination of visible emissions is not likely to be helpful, however.

Another simple form of surveillance, applicable principally to air pollution control, is the remote visual reading. Stack gas plume opacity can be checked this way, and such checks are a common part of most state programs. An opacity reading does not require entry on to the property of the plant. In addition, a trained engineer is not required. An opacity reading can be taken by a "smoke reader," who can be trained in a relatively short time and might even be a volunteer. Typically, readings are taken every quarter hour and are averaged over several hours to obtain the test result. Again, opacity readings provide conclusive evidence of violations only for opacity regulations, and such regulations, in turn, are only imperfect surrogates for limits on particulate emissions.

Other methods of surveillance are available that do not require entry to the source's premises. The most sophisticated of these, involving remote reading instruments capable of estimating concentrations and velocities of particular pollutants in smoke stack plumes, will be discussed further in chapter 3. A variation of this method is the measurement of changes in ambient quality at the property line. In this approach the agency would compare the ambient air quality at the property line on the upwind and downwind sides of the plant. The difference would then be taken to be the contribution of the plant to air quality degradation. While this method is in principle simple, it is of limited usefulness. First, it reveals nothing about emissions from tall stacks unless measurements are made across a wide altitude range, not just at ground level. Second, it does not work well in areas with many plants, because of the difficulty of sorting out the contributions to ambient concentrations of the various sources. Third, even for a single source, property line ambient measurements are imperfectly related to source emissions, because of the stochastic local wind patterns that translate the latter into the former. Because most source emissions are written in terms of mass emission rates, ambient measurements are generally not well suited to the task of enforcement.[3]

Materials balance is yet another method of estimating emissions. In this approach, an accounting is made of the pollutant contained in all inputs and outputs (both product and by-product) of the production process. For example, sulfur emissions from a coal-fired power plant could be estimated by taking the total sulfur content of the coal and subtracting from it the sulfur present in the collected bottom ash and the sulfur captured in any recovery processes.

Materials balance has its own limitations. One is that the agency usually must rely on the plant for at least some of the relevant data (because continuous measurement is necessary to successful application), although spot checks using other surveillance methods are often possible. A second limitation is that this method is not applicable to all pollutants. Stack emissions from combustion processes, for example, include both elemental nitrogen and the pollutants designated NO_x (oxides of nitrogen). There is no way of determining the amount or proportion of each simply from knowledge of the nitrogen in the atmosphere and in the fuel because conversion rates of nitrogen and oxygen to NO_x depend on operating conditions in the fire box. Nor could organic discharges from municipal sewage treatment plants practically be related to "inputs" of food to the community served. More seriously, the ma-

[3] Houston, Texas, and North Carolina are unusual in that they administer some regulations that forbid property-line concentrations from being exceeded, and thus are suitable for property line measurements (Booz-Allen, 1979a, p. IV-3; Booz-Allen, 1979b, appendix E).

terials balance approach demands great precision in measurement when the removal efficiency required by the regulation is very high. This is because the error in estimating the small residual stream (the SO_2 in the stack gas, for example) is greater, in relative terms, than the measurement error in any of the measured streams (input, bottom ash, or recovery plant output).

More direct evidence of source emissions is obtainable only through methods that involve sampling the discharge stream or continuously measuring one or another of its characteristics. The former approach is what is referred to in the rest of this book whenever "monitoring visits," "source tests," or "measurements" are discussed or included as elements of a model. The applicable techniques and their shortcomings are described more fully in chapter 3, but for now it will suffice to point out that these methods are very much more expensive than the simple inspections described at the beginning of this section. Thus, as will be seen below, visits to sources on which discharge measurements are made cost several thousand dollars, while inspections without measurement cost several hundred.

For some pollutants, it is possible to measure concentrations in discharge streams essentially continuously. For example, optical means of continuously measuring SO_2 and NO_x concentrations in stack gases are available and, indeed, required for new sources of air pollution by EPA regulations. More will also be said about these methods in chapter 3. It should be noted, however, that a continuous monitoring instrument, recording its readings, operating close to 100 percent of the time and without deteriorating performance, and under the control of the agency rather than the source, would be a sort of ultimate weapon—providing a benchmark against which the practically available methods of all kinds could be judged.

Enforcement Alternatives

After a source has been monitored and found in violation of an emission standard, the agency has several enforcement responses available, at least in theory. The simplest is the Notice of Violation (NOV), in which the agency informs the owner or operator that a violation has been detected and gives a date before which the violation must end. The source is not actually penalized for violating a regulation, although it may receive some bad publicity.

Other responses by the agency inflict at least some direct injury on the operator. Among the gentlest of these is the conference, in which the agency invites the operator to a meeting to discuss its emission problems. At this meeting the operator outlines what actions will be taken to restore compliance, and the agency sets a deadline by which time compliance will be restored. For all but the smallest firms these meetings are attended by several representatives, such as the plant manager, engineers, and attorneys. The cost of the time spent in such a meeting, therefore, can run into the hundreds or even thousands of dollars.

Another action the agency can take is to require that a source test be performed, at the expense of the plant, to demonstrate that compliance has been restored. As noted, source tests can cost several thousand dollars each, which can be a significant expense for a small plant.

The above actions may not strike the reader as examples of enforcement methods at all. The usual assumption, underlying both the popular conception of enforcement and the formal models surveyed in chapter 4, is that an agency responds to a detected violation of a pollution control regulation by the imposition of legal or administrative sanctions. Indeed, in all states the environmental management agencies are authorized to punish sources with fines or injunctions. But such sanctions are not as easy to use as the common assumption implies.

The first obstacle to the use of legal sanctions is their cumbersomeness. Due process guarantees give the subjects of regulation formidable weapons with which to fight legal actions; and, although they may not "win" in court, violators, by vigorously contesting the imposition of sanctions, can make the effort expensive for the agency. The prospect of such costs may act to discourage agencies from trying to impose fines. Moreover, state and local agencies are often prevented by legislation from pursuing legal remedies directly. Instead, actions must be brought by county prosecutors or the state attorney general. These officers have their own agenda and may not be as interested in pursuing violations of air quality regulations as is the air quality agency.[4] (Melnick in a 1983 book suggests that a similar problem exists at the federal level, as prosecutions come from the Justice Department.)

Application of sanction is also hampered by the difficulty of gathering evidence of noncompliance. As noted in the preceding section, most types of source surveillance are intermittent, thus providing emission data only for isolated days. (Of the exceptions, materials balance measurements provide only data on long-term average emissions, and con-

[4] In San Diego, for example, legal enforcement was hindered until recently by a provision requiring approval by the County Supervisors for any legal actions against a violator (Pacific Environmental Services, 1979). In Houston, a considerable backlog of air pollution cases developed when the city prosecutor was less interested in air quality enforcement than was the city health department (Booz-Allen, 1979a).

tinuous monitors are rarely used for enforcement purposes.) Indeed, only the source test even provides quantitative evidence bearing on violations of mass-emission-rate standards. In bringing an action, therefore, the agency either must be content with prosecuting the violations that were observed on the days that the source was visited or it must ask the court to infer a pattern of violations from the available data. It is possible to draw such an inference because it may meet a "preponderance of evidence" test that is required to sustain a civil action.[5] Nonetheless, it is a leap that courts are unlikely to make without corroborative evidence, such as a history of citizen complaints, a long record of previous violations, or engineering evidence that abatement equipment is inadequately designed or maintained.

Because of the difficulties of administering legal penalties, some states have established a set of administrative penalties for violations of emission standards. These penalties can be assessed without taking the violator to court. Thus, transaction costs are likely to be reduced, although they certainly will not be zero. The penalties may be either arbitrary amounts, related perhaps to source type and violation history, or amounts calculated to equal the economic benefit to the source from noncompliance (its avoided costs). One major difficulty of the latter approach is that it requires calculation of the economic benefits to the source of noncompliance. This calculation is difficult enough when initial compliance is at stake, let alone continuous compliance. At least for initial compliance it is possible to calculate the penalty based on the investment and operating cost of abatement equipment. With continuous compliance, however, calculation of a noncompliance penalty requires the agency to understand the reasons for the violation and the cost to the violator of correction. For violations due to inadequately designed or maintained equipment, these costs are extremely difficult to estimate. Therefore, noncompliance penalties based on the economic benefits of noncompliance appear to have very little practical application in the encouragement of continuous compliance. (For more on the benefits of noncompliance see the discussion of figure 6-2 in chapter 6.)

[5] It almost certainly will not meet the "beyond reasonable doubt" test that is the standard of proof in criminal cases. Although most states provide for criminal as well as civil penalties for violation of air quality standards, criminal penalties are virtually never sought, largely because of the greater burden of proof. But see recent articles, such as in the *New York Times* (Meier, 1985), for evidence that criminal prosecutions may well be emphasized more in the future. EPA has recently issued new civil penalty guidelines (*Environmental Health Letter*, 1984). This policy stresses the dual goals of such penalties—to recapture the costs saved by noncompliance and to provide an extra penalty reflecting the gravity of the environmental results (U.S. EPA 1984).

The agency potentially has other, more severe, methods at its disposal for enforcing compliance. For example, most states require sources to have operating permits that must be renewed periodically—roughly every one to five years, depending on state, source type, and pollutant. Conceivably, permit renewal could be made contingent on emission performance. This is, however, a rather blunt instrument for providing compliance incentives: first, because the consequences of refusing to renew a permit are so dire for the firm and, second, because the duration of the permit is so long that punishment would be postponed, perhaps for years. In any event, permit renewals in most states have been granted automatically and have not been made contingent on previous compliance.

With these catalogs—of surveillance techniques and enforcement options—as background, attention is now turned to the choices actually made by state agencies. Surveillance is dealt with in the next section, followed by enforcement. In each case, initial concentration is on the choices of a single state, New Mexico.[6] The view is then expanded to look at what is known about the choices of other states.

Surveillance Activities in the States

Air Pollution Control Surveillance Activity in New Mexico

New Mexico contains about 120 major stationary air pollution sources (those with potential emissions in excess of 100 tons per year). The exact number fluctuates, principally because of portable asphalt processing plants, which are moved around the state to work on specific paving jobs. These major sources are found in ten different industries as shown in table 2-4. The sources in Bernalillo County (Albuquerque) are listed separately because the local health department has enforcement authority there, while the state Environmental Improvement Agency (EIA) has responsibility elsewhere in the state.

Performance of the tasks of surveillance and enforcement is principally the function of the Enforcement Section within the Air Quality Division of the EIA. Every professional in the Enforcement Section has a technical background, and the section's main functions are to conduct source

[6] In August 1978 Harrington spent three weeks in Santa Fe, New Mexico, in the office of the Environmental Improvement Agency, studying the implementation of air quality regulations in that state. The material in the following section is based on that case study (Harrington, 1981).

Table 2-4.   Industrial Stationary Sources in New Mexico as of April 1, 1978

| Industry | Number of plants | | | Pollutants regulated |
	Bernalillo County	Rest of state	Total	
Natural gas processing	0	38	38[a]	Sulfur dioxide (SO_2), total sulfur emissions
Petroleum refining	0	7	7	Hydrocarbons (HC), 7 other pollutants
Utility steam generators				
Coal-fired	0	3	3	SO_2, nitrogen oxides (NO_x), particulates
Gas-fired	2	8	10	NO_x
Oil-fired	0	2	2	SO_2, NO_x, particulates
Asphalt processing	9	27	36[b]	Particulates
Woodwaste burners	0	10	10	Smoke (opacity)
Sulfuric acid plants	0	2	2	SO_2
Nonferrous smelters	0	2	2	SO_2, particulates
Nonmetallic minerals				
Mica, perlite, pumice	0	7	7	Particulates
Gypsum	1	1	2	Particulates
Cement	1	0	1	Particulates
Gravel	2	0	2	Particulates
Total major sources	15	107	122	

[a]Only eight plants require sulfur recovery units to comply with the regulation.
[b]This is the number of firms; several firms have more than one processing plant.

surveillance and negotiate with plants about achieving compliance.[7] In addition, the Air Quality Division requires approximately 80 percent of the time of one lawyer from the EIA legal section, mostly for enforcement-related matters.

About 75 percent of the resources of the Enforcement Section are devoted to surveillance, while the remainder are devoted to the correction of violations. This breakdown may be a bit misleading because

[7] New Mexico has had some problems with high turnover in its air quality program, principally due to low salaries relative to industry. It is estimated that the average technical employee of the Enforcement Section of the EIA stays only eighteen months. Inasmuch as it takes six to eight months to become fully proficient, each staff member can be expected to produce about a year of useful work before departing. Of the seven professionals in the Enforcement Section, in other words, only four or five are fully contributing members on average.

of the dual function of surveillance. While source surveillance is necessary to find out what sources are doing on a routine basis, it is also necessary in order to find out whether previously discovered violations have been corrected. It did not prove feasible to classify individual surveillance events into these two categories, although it appears evident that the surveillance agenda is sensitive to the compliance status of sources in the state.

Costs of surveillance.   The estimated costs, in 1982 prices, of source tests and inspections are shown in table 2-5. There, the salary, overhead,

Table 2-5.   Estimated Monthly Total and Costs of Source Tests and Inspections in New Mexico

(1982 prices)

	(a) Source Tests
Monthly fixed costs—one source testing team (3 people):	
Salaries, benefits, and overhead	$6,200
Equipment	900[a]
Total	$7,100
Variable costs per test:	
Per diem	$ 450[b]
Vehicle	80[c]
	$ 530
Average cost per test:	
$4,195 at 2 tests per month	
$6,640 at 1.2 tests per month	
	(b) Inspections
Monthly fixed costs:	
Salary, benefits, and overhead	$2,490
Equipment	—
Total	$2,490
Variable costs per inspection:	
Per diem	$ 95
Vehicle	80
Total	$ 175
Average cost per inspection:	
$485 at 8 tests per month	
$795 at 4 tests per month	

Note: Costs inflated from 1978 to 1982 prices using the Producer Price Index.
[a]$51,800 in equipment amortized over five years at 10 percent.
[b]Three days at $50/person/day.
[c]Three hundred miles at $.25/mile.

and prorated equipment cost of maintaining a three-person source testing team is shown as $7,100 per month. To convert this into an average cost per test, the output of a source testing team in tests per month is needed. Depending on the complexity of the source and its distance from Santa Fe, a source test will require as much as two days of travel and three days of actual testing. In addition to time in the field, a significant amount of office work is required, mainly performing calculations and writing up the test results. The objective of the EIA is two tests per month for each source-testing team. Between January 1977 and May 1978 there was only one source testing team, and it conducted twenty-one source tests in seventeen months, for an average of 1.2 source tests per month. (At this rate, the agency can conduct source tests for applicable sources about once every four years.)

If the performance of the source tests is the sole function of the source testing team and 1.2 tests per month are accomplished, then the fixed cost allocated among source tests comes to $5,900 per test. If, however, the EIA objective of two tests per month were in fact met, then the average fixed cost of a source test would be $3,550. Variable costs per test were estimated to be $530 (in 1982 dollars), so that total average costs per test would be between $4,080 and $6,430. This range is roughly corroborated by the data from a survey of state agencies reported in appendix 2-B.

Inspections are considerably less expensive. A cost of $2,490 per month is required to support one engineer-inspector. The EIA standard is four inspections per month per inspector, and inspectors are expected to spend between 50 percent and 100 percent of their time actually doing inspections, which translates into between four and eight inspections per month. With travel and per diem, estimated per test, these alternative assumptions yield average costs between $485 and $795 per inspection.

Surveillance schedule. How does the EIA determine which sources to visit? Although there appears to be only one formal requirement, the agency also uses several rules of thumb to guide the selection of sources. The requirement, from EPA, is that major sources be visited annually as part of the State Implementation Plan. On May 2, 1978, however, approximately 16 percent of the major sources under the EIA's jurisdiction were listed in EPA's Compliance Data System (CDS) as having "unknown compliance status," which means that an inspection had not been conducted within the past year. Table 2-6 shows the average frequency of surveillance for selected categories of sources in New Mexico.

Table 2-6.   Surveillance Frequency of Selected Source Categories in New Mexico

	Number plants in sample	Average visit frequency	Range of visits
		(———— no./yr. ————)	
Four Corners units 4 and 5	1	10.0	—
Asphalt processors	9	1.0	0.4–2.0
Woodwaste burners	5	3.8	2.6–4.0
Nonmetallic mineral processors	7	1.0	0.5–2.0

The EIA also has self-imposed guidelines for certain types of sources. For woodwaste burners, the agency attempts to perform opacity readings at least every other month. In addition, the agency attempts to inspect every portable asphalt plant at every temporary site. Typically, these plants are in operation at a particular site for several weeks to several months. However, it is difficult to obtain an estimate of the EIA's success in visiting all sites, because the material on file reports primarily on the sites the agency knows about and has inspected. Although all asphalt plants in the state are supposed to report to the EIA before they set up temporary facilities, this is not always done.

Surveillance activities are also directed by the receipt of complaints. Based on evidence in the agency files, the EIA does not receive many complaints about industrial stationary source emissions—only about 1.5 to 2 per month—and most have concerned asphalt processors and wood-waste burners.[8] Finally, another consideration guiding the schedule of surveillance is travel. If an inspector must make an overnight trip, he will attempt to visit as many sources as possible in the vicinity.

Once the schedule of plants to be visited has been determined, visits are made in the daylight hours, and the plants are usually notified a week or two in advance. Advance notice is always given when a source test is scheduled. The ostensible reason plants are notified is to ensure that they are operating, so that agency personnel may avoid wasted trips. About the only time the agency conducts a surprise inspection is if there is reason to believe that a violation is occurring and the plant may attempt to conceal that fact.

[8] Interestingly, the complaints in each category were directed primarily at one producer within a category, and complaints for a particular source often came from only one person. Agency and industry sources agreed that these persons were "cranks."

Surveillance by Agencies in Other States

In 1978 the Council on Environmental Quality (CEQ), in conjunction with the EPA, sponsored a research program in air quality implementation. As part of this program the activities of five state and local air quality agencies were examined: New Jersey, North Carolina, Chicago, Houston, and San Diego (See EPA, 1981). At the same time EPA on its own was supporting studies of air quality management in Connecticut, New York, Iowa, and Oregon. In 1979 the National Commission on Air Quality (NCAQ) also sponsored several regional studies of air quality management, in which enforcement was one of the topics of concern. And in 1981 RFF undertook, as part of the study reported here, a mail survey of state agencies involved in the monitoring and enforcement of compliance with air and water pollution control regulations. Taken together, these studies show that the New Mexico findings are broadly typical, though differing in detail, of experience around the country. In particular, it seems that infrequent visits are the rule.

Turning first to the CEQ, EPA, and NCAQ studies it must be admitted that most of the individual state and municipal agency reports suffer from flaws that make their interpretation problematic. Few of them state the total number of sources under the jurisdiction of the air quality agency being examined, and those that do rarely separate major from minor sources or break down the number of sources by source type. Although the total level of surveillance activity (number of visits per year) is given, the lack of data on sources makes it impossible to estimate the frequency of surveillance. Other studies do not indicate the number of violations detected each year. In short, these studies have inadequacies that preclude any systematic analysis, although the clues they offer of what goes on in these agencies tend to support the conclusions of the New Mexico study.

Surveillance methods.   Methods of source surveillance used in these other states are essentially the same as in New Mexico. Primary reliance is placed on engineering inspections. Proof of a violation, however, is possible only for gross emission violations, because the inspection data are of uncertain utility as legal evidence. Several studies suggested that high turnover among inspectors, and consequently a low level of experience, hampers the effectiveness and credibility of the inspections (U.S. EPA, 1981, p. 46).

One noticeable difference between some of these states and New Mexico is that inspections may be unannounced. In New Jersey, which seems to make most use of the unannounced visit, about 75 percent of all inspections are of this type (Environmental Law Institute, 1979, p. 66).[9]

Another difference, at least on paper, is that most of the other states surveyed appeared to require much greater use of continuous monitors, which in New Mexico are mandatory for only one or two sources. In Connecticut, continuous opacity monitors are required on 2,500 fuel-burning sources, and in many of the other states surveyed at least fifty monitors were in place. At present, however, continuous emission monitoring (CEM) data are rarely put to an enforcement-related purpose. Indeed, in Arizona and Colorado it was found that no use at all was being made of the periodic excess emission reports required from new sources, by either the state agencies or the EPA regional offices (Abbey and Harrington, 1981, p. 27). The failure to use CEM data apparently is due, at least in part, to the reluctance of courts to admit those measurement records as evidence of violation. For instance, the Illinois Environmental Protection Agency for years attempted without much success to use CEM data to support enforcement actions and has since abandoned such use (Pacific Environmental Services, 1979, pp. 3–15). Even if CEMs were considered more reliable, their use in enforcement would be hampered by the fact that the vast majority are opacity monitors and, therefore, are badly suited to support violations of mass emission regulations.

Surveillance frequency and cost.   As noted, estimation of surveillance frequency in the states surveyed is made difficult by a lack of information. Where data were reported, they tended to conform to the picture painted for New Mexico.

The most careful of the EPA studies was carried out by the Environmental Law Institute (ELI) for New Jersey, a state that is different from New Mexico in many important respects. In New Jersey, ELI reported 1,177 major facilities (potential emissions of at least 100 tons per year) and 5,500 other facilities subject to the regulations. In 1977 the New Jersey Bureau of Air Pollution Control performed 13,400 source inspections, of which 5,300 were for major facilities. Thus, major facilities were inspected on average 4.5 times per year, while other sources were

---

[9] ELI also asserted that unannounced inspections tended to detect a higher rate of violations than preannounced ones, but the result was not quantified. In any event, even unannounced inspections are usually anticipated by the operator, because most are made just prior to the expiration of the plant's operating permit. Several years ago New Jersey instituted an experimental program of inspections during evenings and weekends. When it found the rate of violations detected to be the same as inspections during normal weekday working hours, the program was discontinued.

inspected about 1.5 times per year. These figures are not out of line with those reported in table 2-6. (In addition, four regional agencies in New Jersey also conduct surveillance, especially of smaller sources.)

ELI estimated the cost of the average field inspection in New Jersey to be $135 (in 1982 dollars), considerably less than the estimate of $500 to $800 for New Mexico, reported above. Part of this difference surely is due to the difference in source density: an inspector in New Mexico spends a large amount of time traveling between widely scattered sources. In addition, the inspections themselves last four hours on average in New Mexico, much longer than the thirty-three to seventy-nine minutes estimated for inspections of various kinds in New Jersey.

For the other agencies surveyed, the data are either sketchy or un-believable. In Houston, well over 7,000 surveillance events per year were reported for 88 major sources and 375 minor ones. These high figures make sense only if they include "drive-by" surveillance—casual observation of odor or visible emissions from car or helicopter. In North Carolina, 7,000 inspections per year are made on average for 930 major and 12,500 minor sources, but the absence of a breakdown of inspections between major and minor sources limits the value of this information. Given the data problems, it is necessary to rely on the assertions of agency personnel, and these generally place the agencies closer to New Mexico than New Jersey. In Oregon, agency representatives assert that 400 sources are visited twice a year, 600 once a year, and 1,000 once in five years. North Carolina leaves it up to the regional offices of the state agency to determine whether sources will be visited quarterly, annually, or biannually. Because in a later chapter a model for monitoring program design is developed that depends on using a source's past history of violations to decide on current surveillance frequency, it is especially interesting to note that in no state was a relationship asserted between current frequency and the previous history of compliance.

Paralleling the New Mexico pattern, agencies in other states do not rely much on source testing to determine continuous compliance. In-deed, some states use source tests hardly at all. In New Jersey, for example, the agency conducts at most sixty-five source tests per year. At that rate, it would take eighteen years to conduct tests only on the major sources in the state. For the most part, however, other agencies appear to use source tests in the same way as New Mexico: about one test for each applicable source every five years.[10]

[10] Not included here are the source tests to establish initial compliance of new sources, which are required by federal new-source regulations within (usually) sixty days of the commencement of operations. Generally these source tests are arranged and paid for by the companies themselves.

Complaints.   The most striking difference between New Mexico and the other programs surveyed by EPA is in the role of complaints. As noted, complaints in New Mexico are uncommon: only about fifteen per year for major sources. In contrast, the number of complaints reported by some other agencies is as follows (data are from U.S. EPA, 1981, pp. 69–72):

Agency	Estimated number of complaints per year
Chicago	1,500
Houston	2,000
Connecticut	1,000
New Jersey	2,700
North Carolina	500
Oregon	1,000
Iowa	300

Thus, in highly industrialized areas complaints are received at a rate 100 times greater than in New Mexico. Only a part of this difference is accounted for by the difference in the number of sources. The rest is probably attributable to differences in population density. With large numbers of people living in close proximity to the sources of pollution, it is perhaps inevitable that complaints are more frequent. Most of the sources in New Mexico are located in isolated rural areas. (Recall that the one large metropolitan area in the state—Albuquerque—is not under the jurisdiction of the EIA.)

Because virtually all complaint reports are investigated, complaints are an important determinant of the surveillance schedule. However, relatively few complaints actually result in the issuance of NOVs—about 10 percent in Houston, for example.[11]

Results from a Survey of State Officials

At about the same time that the EPA and CEQ study reports were being circulated, a mail survey of state officials was undertaken by RFF in an attempt to answer several basic questions that would help to determine the direction of this study of monitoring and enforcement. The survey is described in some detail in appendix 2-B, and some of the results were already used in the first section of this chapter. The

[11] For obvious reasons virtually all complaints concern either visible emissions or odors. Odors comprise more than 50 percent of the complaints in Connecticut, 61 percent in Houston, 76 percent in New Jersey, and 33 percent in Chicago.

following observations, however, are relevant to the discussion of state practice.

1. Self-monitoring is by far the dominant choice for dealing with what might be called primary surveillance, that is, the routine checking of compliance status. This is true not only for major new sources, for which EPA regulations require it, but also, in most states, for all major sources and many smaller ones as well. Indeed, most agencies reported that there were no (or very few) sources for which they provided primary monitoring. Only one state reported not using self-monitoring at all for existing sources.

2. Auditing, or checking of self-monitoring sources by the agency, is most commonly conducted on an annual basis, but the reported frequency of audits ranged from once each month to once every five years. When the frequencies were averaged separately for air and water pollution and source size, there appeared to be differences between air and water, and between large and small sources within each medium. Statistical tests of significance of these apparent differences, however, revealed that almost all could have resulted from chance occurrences in the pattern of responses (appendix 2-B). However, it does appear that large sources are audited more frequently than small sources (almost twice each year on average versus less than once each year for air pollution, for example); and sources of water pollution are audited more frequently than sources of air pollution, controlling for source size.

3. The reported costs of audit visits vary as widely as their frequency—from roughly $100 in some states when discharge measurements are not made to several thousands of dollars in other states when such measurements are made. The average of reported cost of an audit visit (presumably in 1982 dollars) is in the range $150 to $300 when no measurement is involved (an inspection) and $1,000 to $1,500 when source measurement is involved. The cost difference between measurement and nonmeasurement visits is consistent with the estimates from New Mexico reported above. But the figures do suggest that New Mexico's costs are higher than average.

4. Thus, it seems that the cost of current practice, with about two audit visits per year to large sources, is on the order of $2,000 to $2,500 per source. This may be thought of as the cost of adding a new source to the average agency's responsibility.

5. State agency budgets and manpower allocations varied widely and apparently not entirely in accordance with the number of sources to be checked. However, the questionnaire did not ask sufficiently detailed questions to pick up differences in accounting and assignment rules, so much of the variation may simply result from differences in details such as where enforcement lawyers are included in budgets and how to account for equipment. Using agency budget totals and estimated visits per year to sources, ranges of average costs per auditing or monitoring visit were calculated. These are broadly consistent with the reported costs per visit (appendix 2-B).

Enforcement Actions

As explained above, the word "enforcement" is used here to denote the activities that follow detection of a violation. Notices, mandatory meetings, required additional source tests, and fines are all examples of enforcement actions. Their goal is in principle a dual one: to bring a violator back into compliance and to dissuade that source and others from future violations. As will be seen, however, most of the observable enforcement effort in states today appears to be aimed at the first of these goals. This concentration on correction and apparent lack of interest in future incentives raises intriguing questions for later consideration.

Enforcement in New Mexico

New Mexico relies heavily on voluntary compliance in its approach to enforcement. In fact, the state's air quality legislation requires that the EIA give the offending source a chance to come back into compliance voluntarily.

Procedurally, voluntary compliance works as follows. When, as a result of a source test or inspection, a plant is found to be in violation, the agency sends the owner or manager a notice of violation, asking for a reply within two (or sometimes four) weeks to let the agency know what plans are being made to come back into compliance. The owner may respond in one of several ways.

First, the company may reply that it was experiencing an upset condition that caused the equipment to perform improperly. The agency will then reschedule the inspection and inform the operator to submit an 801 report in the future. (Pursuant to Air Quality Regulation 801, a plant is supposed to notify the EIA within twenty-four hours of any upset or startup condition that leads to excess emissions.)

Another possible response to the violation notice is to report that the control equipment was in need of repair. Generally, at the same time the source would report that arrangements had been made to get the equipment repaired and give a date when it expected to be back in compliance. If this date is not too far into the future and if the source has not been a persistent violator in the past, the EIA generally will accept this solution; it will either schedule an inspection or source test for shortly after compliance has presumably been achieved or else order the plant to conduct a source test and report on the results. Often, however, the owner may not know why he was having problems, or, if he does know, the condition may not be easily correctable. In that case, the agency and the source may negotiate either an "assurance of discontinuance" or a variance. Each is a type of agreement between the agency and the source in which the agency agrees not to enforce the regulation as long as the source meets certain stipulations (Harrington, 1981, chapter 4).

The final possibility for response is defiance. The owner may simply decline to answer the letter from the EIA notifying him of a violation, or he may deny the validity of the test result or inspection, or even of the regulation itself. Defiance is rarely if ever the first response of a source; usually it comes after a period of negotiation between the source and the agency that results in an impasse.

If, in the view of the Enforcement Section, the plant is not complying voluntarily, a letter is sent from the Legal Section of the EIA. This letter informs the plant that if it cannot show evidence of progress toward compliance within a certain time limit, then legal proceedings will be initiated. If this letter does not produce results, then the EIA will eventually resort to court action. The case is turned over to the state attorney general, who may decide to file a complaint. It is extremely rare for a case to get this far without resolution, and the elapsed time before this step is reached is measured in years, not months.

On the evidence, it would appear that the voluntary compliance approach used by the EIA is almost always ultimately successful in bringing a source back into compliance. In fact, the vast majority of sources return to compliance readily after first being informed of a violation. While sources generally are brought back into compliance, it is not clear that voluntary compliance provides any compliance incentives until after a violation has been discovered because a plant knows that it will have a chance to return to compliance before sanctions are imposed. Thus, infrequency of surveillance and the fact that few violations are corrected immediately together create a situation consistent with the fragmentary evidence given in chapter 1 showing that rates of continuing compliance are only fair.

Enforcement in Other States

With few exceptions the enforcement behavior of the state and local agencies surveyed by the EPA and CEQ resembled that of the New Mexico EIA.[12] For one thing, discovery of violations rarely led to court cases. In North Carolina between 1975 and 1978 an average of 180 NOVs each year were issued, of which only 5 resulted in court penalties (Booz-Allen, 1979a, pp. V-39 and V-48). In San Diego over the same period, of about 300 to 400 NOVs issued each year, court cases were filed for only about 10, although nearly all have resulted in convictions (Pacific Environmental Services, 1978, pp. 3-31 and 3-36). In Houston more than 1,800 NOVs were issued in 1977, of which only 8 were referred to the city attorney. In 1978 the referrals jumped to 48, a substantial increase but still a small fraction of the total NOVs issued (Booz-Allen, 1979b, pp. IV-22 and appendix G). Thus, the vast majority of violations did not result in any administrative or civil penalty at all.

Second, where penalties are imposed, they are very small. The *total* fines imposed each year in a subsample of the state and local agencies surveyed are as follows: San Diego, $4,000 (in 1978); North Carolina, $3,750 (avg. 1975–78); Oregon, $11,850 (1978); and Houston, $3,850 (1978); (U.S. EPA, 1981, p. 133; Booz-Allen, 1979a, p. V-47; U.S. EPA, 1981, p. 104; and Booz-Allen, 1979b, appendix G). Average penalties are thus a few hundred dollars. It seems clear that fines of this magnitude cannot by themselves provide any serious incentive to compliance.

It is interesting to note that *authority* to levy much higher penalties is often available to the agencies. For example, in New Jersey, the Bureau of Air Quality Protection has the authority to issue, upon discovery of a violation, a Notice of Prosecution and Offer of Penalty Settlement (Environmental Law Institute, 1979). The penalty depends on type of source but generally increases with successive violations. Upon receipt of this notice, the violator may either pay the penalty, negotiate with the agency for a reduced penalty, or refuse the offer of settlement, in which case the matter is referred to the attorney general for prosecution. If the penalty is accepted, the violator may apply later for a rebate of a portion of it. Note that the operator always has the

---

[12] The RFF survey contained no questions on enforcement practices.

Table 2-7. State Enforcement Activity

(annual averages, 1978–83)

State	NOVs issued (1)	Civil actions brought (2)	Number of penalties assessed (3)	Penalties ($) Total penalties assessed (4)	Penalties ($) Average size of penalties assessed (5) = (4)/(3)	Penalties ($) Average penalty per NOV (6) = (4)/(1)
Colorado	124	3.6	0.5	60	120	0.5
Connecticut	800	2.3	21.5	7,800	363[a]	9.8
Indiana	59[b]	NA	21.0	85,000	4,050[c]	1440
Kentucky	194	5.2	5.2	13,100	2,520[d]	68
Massachusetts	NA	0	0	0	0	0
Minnesota	41	NA	10	109,000	10,900	2660
Nebraska	59[e]	NA	0.2	400	200	0.07
Nevada	31.5	0.3	2.3	105	45	3.3
New Jersey	1,167	NA	350	500,000	1,430	428
Oregon	197	NA	30.7	21,600	705	110
Pennsylvania	NA	NA	176[f]	260,000	1,480	NA
Rhode Island	5	7.2	0	0	0	0
South Carolina	68[g]	5.5	2.2	53,400	24,250[h]	785
South Dakota	17	1.2	0.3	300	1,000	20
Tennessee	193[i]	8.4	0	0	0	0
Virginia	161	7.8	3	600	200	3.8
Wisconsin	80.5[j]	13.5	7.7	61,200	7,951	760

Note: NOV = Notice of Violation, NA = not available.
Source: Resources for the Future survey.

[a]Refers to amounts assessed; over the same period actual collections were 62 percent of assessments.

[b]Includes both NOVs and Compliance Orders.

[c]Excludes one penalty of $415,000, which was cancelled when company bought equivalent amount of air pollution equipment.

[d]Excludes performance bonds (two required for total of $45,000 in last five years).

[e]Excludes Lincoln and Omaha.

[f]Only data from the last quarter of 1983 were readily available. This figure is an extrapolation to an annual average.

[g]No NOVs were issued before April 18, 1983. From July 1, 1983, to March 31, 1984, fifty-one NOVs were issued.

[h]Excludes one fine of $1,700 per month until compliance was restored and one fine of $250,000 dropped in consideration of a donation of a like sum to a technical college.

[i]Does not include NOVs from Continuous Monitoring Data, which were extremely numerous.

[j]1980–83 only.

41

option of getting its case into the courts. That most (85 percent) do not probably indicates that the size of the actual penalties (under $1,000 per day) does not justify undergoing the risk of a greater penalty as a result of the court case (Environmental Law Institute, 1979, p. 94).

A different version of the administrative penalty is found in Connecticut. The Connecticut Enforcement Act of 1973 authorized state air quality officials, upon discovery of a violation, to levy a "noncompliance penalty" equal to the economic benefit of noncompliance. Few such penalties have actually been assessed, although state officials argue that the mere existence of an administrative penalty has provided incentives for compliance. They claim that warning violators of potential liability for administrative penalties has reduced compliance delays by 30 to 40 percent (Connecticut Enforcement Project, 1975, vol. 2).

A third point is that the vast majority of NOVs, as well as penalties, involve violations of opacity or odor regulations, reporting requirements, and other easily proved conditions. Few penalties are assessed for violations of mass emission regulations. In 1978, for example, Arizona issued twenty-eight NOVs, of which only one was for a mass emission violation (Abbey and Harrington, 1981, p. 35). The others were for violations of opacity regulations, permit conditions, reporting requirements, and the like. In Houston, of the ninety-three cases referred to the city attorney between 1974 and 1978, only sixteen were for violations of mass emission rates (Booz-Allen, 1979b, appendix G). This is hardly surprising, given the infrequency of source-testing visits.

In order to get more recent information on state enforcement activity, RFF in 1984 informally surveyed all fifty state air quality agencies on their enforcement activity in the past five years. Replies were received from seventeen states, and the responses are summarized in table 2-7. This table includes fines for violations of both initial and continuing compliance, so it overstates somewhat the use of penalties for the latter.

The most interesting information in the table is found in the rightmost column, which in effect is what a source can expect to pay if a violation is discovered. In eight states it is less than $10; and only in Indiana, Minnesota, New Jersey, South Carolina, and Wisconsin is the expected cost of a violation greater than $400. Thus, states can be divided into two groups: a large group that makes virtually no use of penalties to enforce air quality regulations and a much smaller group that levies apparently sizable penalties. (Data on total NOVs in Pennsylvania are not available, but the large number of penalties and their large average size suggest that this state should be included among the "punitive" states.) The punitive group can be further divided into a set that levies a small number of very large penalties (represented here by Indiana,

Minnesota, South Carolina, and Wisconsin) and one that levies a large number of smaller penalties (New Jersey and Pennsylvania). In this regard it is interesting to note that Indiana and Minnesota report an anomalously small number of NOVs, given their size. This could be either because a different definition of NOVs is used in these states or because the record of heavy penalties has acted as a deterrent.

Thanks to the ELI study there is more detail on New Jersey (ELI, 1979). In recent years the state agency has reported detection of about 1,200 violations each year. About 400 of these were minor violations or equipment upsets, which the agency deals with informally. For the remaining 800 cases an administrative order was issued, which required the source to return to compliance within one to three weeks (depending on the type of source or the violation) or face the possibility of being shut down. For 250 of the administrative order cases, a notice of prosecution was issued. For these cases the agency levied an administrative penalty that varied in severity with the number of previous violations (within the permitting period) and also with the type of regulation. For violations of mass emission rates, for example, the first violation typically resulted in a compliance order, the second in a $2,500-per-day fine, and the third was referred to the state attorney general for civil prosecution (ELI, 1979, p. 82). Though exact figures are unavailable, this system appears to result in several hundred thousand dollars in fines every year. Compared with those of other states, this system is practically draconian. Compared with the millions required to operate pollution abatement equipment in New Jersey, however, the fines appear rather small. Thus, even in New Jersey it appears that penalties for noncompliance are unlikely, in themselves, to be sufficient to induce continuous compliance.

Conclusion

In this chapter the practices of state environmental agencies in enforcing continuous compliance with air and water quality regulations have been examined, and a number of general points have been made. The determination of the compliance status of most sources operating under limits on emissions per unit time is quite expensive. Source measurement provides estimates of mass emissions but is expensive and not applicable to every class of source. Less expensive methods do not provide enough information to determine compliance with any confidence. Moreover, even the less expensive methods (such as inspections and opacity readings) are infrequently used for any particular source. Indeed, many sources are not even visited as often as once a year, although those that are visited most often tend to be sources with especially high emissions or that have triggered citizen complaints.

When violations are detected, the offending source is almost always given the opportunity to return to compliance without any sanction imposed for past violations. Given the difficulties of surveillance, this use of the voluntary compliance technique is hardly surprising. As noted above, however, voluntary compliance does not appear to provide sources with a very strong incentive to maintain continuous compliance, at least on first consideration. This proposition will be examined in more detail in chapter 4.

APPENDIX 2-A

The Text of an "Air Pollution Control Permit to Construct" Issued by the North Dakota State Department of Health

Pursuant to the Air Pollution Control Regulations of the State of North Dakota and in reliance on statements and representations contained in the Permit to Construct application and supplemental information supplied to this Department, the State Department of Health hereby grants _____ a Permit to Construct a _____ plant in Sections _____ and _____ , Township _____ , Range _____ in _____ . This Permit to Construct is subject to all applicable rules, regulations and orders now or hereafter in effect of the North Dakota State Department of Health and to the conditions specified below:

1. This permit shall in no way permit or authorize the creation and/or maintenance of a public nuisance or a danger to public health or safety.

2. All reasonable precautions shall be taken to prevent and/or minimize fugitive dust emissions during construction. Thirty (30) days prior to on-site construction the owners shall submit a detailed plan of their proposed actions to this Department for review and approval.

3. The following emission monitoring equipment shall be installed on all major fuel burning sources. This equipment shall be calibrated, main-

tained and operated by the owner or operator. This equipment includes:

(a) A continuous monitoring and recording system to measure the opacity of emissions.

(b) A continuous monitoring and recording system to measure sulfur dioxide emissions.

(c) A continuous monitoring and recording system to measure nitrogen oxide emissions.

(d) A continuous monitoring and recording system to measure either oxygen or carbon dioxide in the flue gases.

(e) The continuous monitoring and recording system shall be capable of demonstrating compliance with applicable standards and the emission limits specified in Condition 9 of this Permit to Construct.

These items are to be installed and operated as specified in Section 12.111 and 12.401 of Regulation R23-25-12 of the North Dakota Air Pollution Regulations.

4. (a) Continuous monitoring equipment to measure and record the concentration of total sulfur in the Stretford Unit tail gas prior to incineration of the tail gas in the superheater furnaces shall be installed, calibrated, maintained and operated by the owner or operator. The results of this monitoring shall be such that the contribution of SO_2 from the Stretford Unit tail gases to the superheater furnace flue gases may be readily determined.

(b) Continuous monitoring equipment to measure and record the concentration of total sulfur in the Shift Catalyst Regeneration Unit off gases shall be installed, calibrated, maintained and operated by the owner or operator. The results of this monitoring shall be such that the contribution of SO_2 from the Shift Catalyst Regeneration Unit off gases to the main stack emissions may be readily determined.

5. This Department shall receive written notification of the following:

(a) The anticipated date of initial start-up, not more than 60 days or less than 30 days prior to such date.

(b) The actual date of initial start-up within 15 days after such date.

(c) The date upon which demonstration of the continuous monitoring system performance commences, postmarked not less than 30 days prior to such date.

6. Following completion of construction and within 60 days after

achieving the maximum production rate at which the facility will be operated, but not later than 180 days after the initial start-up, the owner or operator shall conduct performance tests of emission sources specified by this Department. Test methods used shall be those set forth in the North Dakota Air Pollution Control Regulations or equivalent methods approved by the Department. Performance test guidelines shall be obtained from this Department prior to developing testing procedures, to insure they will comply with Department requirements. This Department shall be notified 30 days prior to the date of the tests to allow for a pretest meeting with the Department and to afford the Department the opportunity to have an observer present. The test results shall be submitted to the Department for review and approval.

7. Sampling ports shall be provided downstream of all emission control devices and in any flue, conduit, duct, stack or chimney arranged to conduct emissions to the ambient air. The ports shall be located to allow for reliable sampling and shall be adequate for test methods applicable to the facility. Safe sampling platforms and safe access to the platforms shall be provided. Plans and specifications showing the size and location of the ports, platforms and utilities shall be submitted to the Department for review and approval.

8. An ambient air quality and meteorological monitoring program shall be undertaken. The program shall be designed to determine the ambient background concentrations for designated pollutants in the vicinity of the plant site before initial start-up and to verify compliance with ambient air quality standards after the plant is in operation. A site specific air quality dispersion model for the _____ shall be developed as a part of this program to correlate fuel and plant operational conditions, local meteorology and coincident ground level ambient air quality concentrations to verify the stack design parameters coupled with the emission control systems' ability to satisfy ambient air quality standards. The principal pollutants to be monitored shall include, as a minimum, total suspended particulates, total sulfur, sulfur dioxide, nitrogen oxides (NO, NO_2, NO_x), suspended sulfates, suspended nitrates, suspended fluorides, suspended particulate pH, sulfuric acid mist, ozone, hydrogen sulfide, hydrocarbons and sulfation rate. This monitoring shall begin a minimum of 1 year prior to initial start-up and continue for a minimum of 2 years after normal operations have been established. Cessation of the monitoring program will be based on the Department's evaluation of the data collected. Prior to final commitment for the monitoring program a detailed description, including sampling and analysis methods, location of sampling sites, instrumentation spec-

ifications and data handling shall be submitted to the Department for review and approval.

9. Construction shall be in accordance with the plans and specifications furnished and representations made to this Department. The maximum permitted emission rates for the various air contaminants are based upon those specified in the permit to construct application when less than applicable State emission rate standards. The emissions of the major contaminants are limited as follows: [See table 2-1 in the chapter text.] The Department shall be notified in advance of any significant deviations from the specifications furnished to allow time for review and approval. The issuance of this permit to construct may be suspended or revoked if the Department determines that a significant deviation from the plans and specifications furnished has been or is about to be made without prior approval from the Department.

10. Plans and specifications for all pollution control devices, including the electrostatic precipitators, wet scrubbers and fabric filter baghouses, shall be submitted to the Department, for review and approval, thirty (30) days prior to the start of fabrication. Design calculations supporting the efficiency claims of the control devices and showing that each respective emission source will comply with applicable State and Federal standards shall accompany the plans and specifications. A timetable listing the dates this material will be submitted to the Department shall be included with the first monthly construction progress report.

11. Plans and specifications for all coal conveying and handling equipment and the ash handling system shall be submitted to the Department, for review and approval, thirty (30) days prior to the start of fabrication. A timetable listing the dates of material will be submitted to the Department shall be included with the first monthly construction progress report.

12. Periodic monthly reports on construction progress shall be filed by the owner with the Department not later than 15 days after each month. The first report shall be submitted forty-five days after the start of construction. Any revisions to the project schedule shall be submitted with the report.

13. Any violation of a condition issued as part of this permit to construct, as well as construction which proceeds in variance with any information submitted in the application, is regarded as a violation of construction authority and is subject to enforcement action.

14. This permit shall in no way permit or authorize the discharge of air contaminants which cause an odor, outside of the plant boundary, that is objectionable to individuals of ordinary sensibility.

15. A site specific monitoring program for the _____ shall be developed to verify the odor model and demonstrate compliance with Condition No. 14 by correlating fuel and plant operational conditions, local meteorology and coincident ground level odor concentrations. This monitoring shall begin upon initial start-up and continue after normal operations have been established. Cessation of the monitoring program will be based on the Department's evaluation of the data collected.

16. Detailed plans and specifications as well as additional supporting technical data and information, demonstrating that all technological advances developed following the issuance of this permit have been examined and investigated and those found to be practicable and feasible have been incorporated into the design for Phase _____ construction, shall be submitted to the Department for review and approval not later than 180 days prior to proposed initiation of construction of Phase _____ of the _____ .

17. This Permit to Construct is issued in reliance upon the accuracy and completeness of the information set forth in the application. Notwithstanding the tentative nature of this information, the conditions of this permit herein become, upon the effective date of this permit, enforceable by the Department pursuant to any remedies it now has, or may in the future have, under the North Dakota Air Pollution Control Law, NDCC Chapter 23-25. Each and every condition of this permit is a material part thereof, and is not severable.

_____ has reviewed the terms and conditions of this Permit to Construct and finds them to be reasonable in light of the information and representations that have been made available to the North Dakota State Department of Health.

This Permit to Construct for the _____ plant shall be deemed acceptable by _____ and becomes fully effective on the date that this Permit to Construct is signed by a corporate official and returned to the North Dakota State Department of Health.

Date _____   North Dakota State Department of Health
                           Air Pollution Control Program

                           By: _____
                           Title: *Chief, Environmental Control*

Date _____

                           By: _____
                           Title: *Executive Vice President*

APPENDIX 2-B

The RFF Survey of State Agency Surveillance Activities and Practices

The procedures used in the survey were straightforward. The questionnaire was aimed at six areas of inquiry:

- the size of the responding agency's surveillance task (numbers of sources by type) and whether that task was for air or water pollution, or both;
- the nature of the discharge standards written into the permits to be enforced;
- the extent to which self-monitoring by dischargers is required and what specifications are placed on its performance;
- the nature, frequency, and cost of audits performed by the agency on self-monitoring sources;
- the extent to which the agency provides the primary (only) monitoring (discharge measurements) for some sources, how much effort it expends on that activity, and how primary monitoring differs from auditing of self-monitored sources—if indeed it does.
- the sizes of agency surveillance staffs and budgets.

After checking the survey form with several people involved in source surveillance in eastern states, it was mailed in the spring of 1982 to 150 addresses in the fifty states and the District of Columbia, using addresses from the 1982 edition of the National Wildlife Federation's *Conservation Directory*. Sixty-three responses containing some usable data from forty-two states were received, though not every response contained answers to all questions, even after allowing for the distinction between air and water quality agencies. The results of the survey fall under seven headings:

- response rates overall and in specific question areas;
- sizes of agency responsibilities;
- types of standards imposed on sources (covered in text, not appendix);
- extent of self-monitoring and the nature of the requirements imposed;
- nature, frequency, and costs of audits of self-monitored sources;
- similar information on primary monitoring visits;
- manpower and overall budget allocations, and implied costs per monitoring visit.

Number and Content of Responses

Sixty-three more or less complete questionnaires were received from forty-two states. The twenty-one "extra" responses for the most part reflect that separate agencies monitor air and water pollution in most states. (For many states only an air or only a water agency response was received. For twenty-one states there was coverage of both air and water practices. For a few states multiple responses reflecting regional subagencies were received.) The number of responses (not all separate) covering air pollution monitoring was thirty from twenty-seven states; for water, forty-four from thirty-six states.

The numbers of agencies attempting at least some answer to the key questions are shown in table 2B-1.

Thus, 82 percent of the fifty states plus the District of Columbia responded in some fashion, and for about 65 percent some budget data were received. This was gratifying, especially considering a confusion over questionnaire versions, the current financial strains reported by state governments, and the small return on the investment of time that could reasonably be expected by any responding agency or individuals.[1]

Numbers and Types of Sources Monitored

The number of sources for which agencies are responsible varies tremendously across the states, as one would expect. Taking averages over the agencies providing enough information to allow calculation of a source total produces for air, 4,550 dischargers per average state agency, and for water, 1,770. But these numbers have very little meaning when the ranges underlying them are recognized.

Table 2B-1. Summary of Number of Responses by Question Category

Question category	States	Responses
Character of permit terms	All respondents	
Number enclosing sample permits or similar documents	14	15
Audit frequency	41	61
Audit cost	22	29
Primary monitoring visit frequency	19	21
Manpower data	34	56
Budget data	33	52

[1] One questionnaire was returned blank except for a note pointing out that under current budget tightness no one had time to fill out forms for inquisitive researchers. Several other returns were essentially blank and are not counted as "responses."

One possibility is that the top and bottom extremes, for example, the tenth and ninetieth percentiles of the responses, are in fact based on a different understanding of the question or different definitions of "responsibility" or "source," and might therefore reasonably be eliminated.

	Range in reported number of sources for which agencies are responsible	
	Low	*High*
Air	22	72,800
Water	32	5,100

If this is done and the means recalculated, for air the new mean is 1,325 sources, a large reduction, while for water it is 1,670, a very small change. The remaining ranges are still large, however:

	Range in reported number of sources for which agencies are responsible	
	Low	*High*
Air	40	8,140
Water	220	3,920

Similar, and similarly uninstructive, variation is found in the reported numbers of particular source types of both air and water pollution for which the agencies are responsible. For completeness, but with a caution not to attach any particular importance to them, the means and extremes of these reported numbers are tabulated in table 2B-2.

One particularly interesting finding, however, concerns the proportion of total sources required to do self-monitoring. This could be determined for twenty-two air pollution monitoring agencies and thirty-three water pollution monitoring agencies. (The determination was made in a few cases on the basis of an explicit statement that "all" or "no" sources were required to monitor themselves. In all other cases where a determination was possible, it was based on reported numbers of self-monitored sources and of total sources for which the agency is responsible.) For air pollution sources, a self-reporting requirement is far from rare. Of the twenty-two reporting agencies, four require all air pollution sources to self-monitor. Overall, the average proportion of air pollution sources required to self-monitor is about 0.28. But on the water pollution control side, self-monitoring is practically universal. Of the thirty-three reporting agencies, two-thirds, or twenty-two, require all water pollution sources

Table 2B-2.   Number and Type of Sources for Which Agencies Are
Responsible

Sources	Means of reported numbers[a]	Extremes of reported ranges
Air		
Industrial plants	2,514	7–37,580
Electric generation plants	19	2–68
Incinerators	581	1–4,800
Other	4,217	2–30,400
Water		
Industrial plants	623	14–2,500
Municipal sewage treatment plants	578	26–3,300
Feedlots	526	0–2,500
Other	229	4–2,000

[a]Means of explicitly reported numbers, including explicit zeroes; but blanks were not interpreted as zeroes.

to do some self-monitoring. And the overall average proportion of sources required to self monitor is 0.84.[2]

Self-Monitoring

As already described, self-monitoring of discharges is required of all dischargers by a number of states and of at least some dischargers by all but one of the responding states. Some specific questions about the content of these requirements were also of interest—for example, whether particular makes of equipment are required or only certain performance standards; what calculations, if any, must be done on the raw data by the source; and what information has to be sent to the responsible agency.

The responses to the questions on these matters are summarized in table 2B-3. A few of these are worth detailed scrutiny. First, very few agencies specify monitoring-equipment make—and, indeed, it is possible that those few that indicated they did specify make might have thought the question meant generic type. On the other hand, the requirement that the equipment record automatically is very common and would certainly serve to cut down on opportunities for tampering with

---

[2] The difference between the air and water proportions (0.28 and 0.84) is statistically significant at the 2 percent level; that is, there is less than a 2 percent chance that the difference in the proportions is spurious.

Table 2B-3. Requirements Imposed on Self-Monitoring Sources

Category	Total	Air	Water
Equipment			
Total responses	57	26	31
Proportion specifying make	0.07	0.12	0.03
Proportion specifying precision[a]	0.47	0.65	0.32
Proportion specifying accuracy[a]	0.60	0.77	0.45
Proportion specifying automatic recording	0.81	0.81	0.81
Proportion with other equipment specifications[b]	0.26	0.35	0.19
Calculations by sources			
Total responses	66	25	41
Proportion requiring averaging over time	0.95	0.96	0.95
Proportion requiring standard deviations[a]	0.14	0.24	0.07
Proportion requiring moving sums[a]	0.20	0.32	0.12
Proportion requiring other calculations[c]	0.21	0.12	0.27
Reporting by sources			
Total responses	68	26	42
Proportion requiring raw data record[a]	0.53	0.31	0.67
Proportion requiring averages over time[a]	0.78	0.58	0.90
Proportion requiring violations over time[a]	0.82	1.00	0.71
Proportion requiring other summary statistics[d]	0.21	0.15	0.24

[a]The differences between the proportions for air and water monitoring are significant at 5 percent for these items. (See chapter 5 for definitions of accuracy and precision.)

[b]Other specifications for air monitoring equipment include: that the equipment meet EPA specification (four); that it meet other, unlisted, specifications (two); that it be equipped with an audio alarm (one); and two other miscellaneous requirements. For water monitoring equipment, meeting EPA specifications was a requirement listed on three forms, a method-of-analysis requirement was mentioned on two, and a calibration requirement on one. No significance test was performed on the difference between air and water proportions.

[c]Other calculation requirements listed by respondents included, for air: violations (two) and "emissions profile" (one). For water: six responses listed some version of a requirement that highest and lowest values be found; three required some other version of the central tendency, such as the geometric mean; one required a percentage removal; and one a calculation of the percentage of time over standards. No significance test was performed.

[d]Other reporting requirements imposed include: for air—fuel sulfur content; for water—measures of highest values; various indicators of plant operation and performance such as "pond freeboard," "haulage records," "loading values," and "operational data" in general. No significance test was performed.

records. Performance specifications are more commonly imposed on air than on water pollution sources.[3]

When it comes to calculations the only common requirement is averaging over time. This is consistent with responses to other questions that indicate a general absence of statistical tests for violations. The

[3] We asked specifically about precision (the size of random errors of reported values around true values) and accuracy (the ability of the equipment to find the true values without bias). See chapter 4.

responses listed under the "other" category did contain a number of references to statistics measuring the extremes of recorded discharges, such as the maximum and minimum and the tenth and ninetieth percentiles. Such measures may be more commonly required than the responses indicate, because no simple checkoff opportunity listing them was provided.

The reporting requirement most often imposed on sources is that "violations" be identified. This is done by every agency with self-monitored air pollution sources responding to this set of questions.[4] It is significantly less common for water pollution control agencies to impose it. The other two listed reporting requirements, raw data and averages over time, are imposed more commonly on water pollution than on air pollution sources, and the latter requirement is very nearly universal among the responding water pollution control agencies.

Auditing Self-Monitoring Sources

Because the requirement for self-monitoring appears to be widespread, the question of how, and how often, the agencies check up on these sources becomes important. Agency practices in this regard are summarized by the figures in tables 2B-4 through 2B-6.

Table 2B-4.   Auditing Self-Monitored Sources: Frequency of Audit Visits

| | Type of source | | | |
| | Air pollution | | Water pollution | |
Type of visit	Large	Small	Large	Small
Without measurement				
Responses	26	15	41	35
Mean (times/yr.)	1.70	0.88	3.10	1.44
Standard deviation	1.19	0.64	3.21	1.71
Including measurement				
Responses	18	10	34	26
Mean (times/yr.)	1.43	0.89	2.41	1.39
Standard deviation	1.26	0.81	2.91	1.72

[4] This finding raises an intriguing question when coupled with those from the calculation section: How are violations to be identified when apparently none of the statistical apparatus necessary to separate out true violations is required? The answer must lie in rules of thumb, some of which are mentioned elsewhere in individual survey responses. An example of one such rule of thumb is: if the second highest emission measured (over whatever averaging period) is greater than the standard, then there is a violation.

In the first table, the following patterns appear to be present:

1. For both air and water pollution sources, large sources are audited more often than small sources. Large air pollution sources are audited on average (across the reporting agencies) a bit less than twice each year. Small air pollution sources are audited less than once each year on average. For water pollution sources, the corresponding figures are a little more often than three times per year and a little less often than one and one-half times per year (or about once every eight months).
2. For both air and water pollution sources, not all audit visits seem to include actual discharge measurement, though this difference is only pronounced for large water pollution sources.
3. Water pollution sources, large and small, seem to be audited more often than air pollution sources in the corresponding size classes.

When these apparent patterns are examined statistically, however, only one of the differences turns out to be significant at the 5 percent level, that is, to have a probability less than 5 percent of having been due to random variation in the data. The difference between the audit frequencies for large and small sources of water pollution passes the significance test. Thus, the apparent lessons of the table of audit frequencies must be interpreted rather as artifacts of very large variations in state schedules.

Table 2B-5, on audit conduct, contents, and the use of data developed by audits, shows that a large fraction of responding agencies announces audit visits, either always or sometimes. Almost all audits involve inspections of records and equipment. Roughly consistent with the above observations about frequency of discharge measurement, the responses to a separate question on whether or not audits involve measurement says than 90 percent *of the agencies* consider that they do. Most agencies use audit data for purposes both of identifying potential trouble and identifying actual violations.

Table 2B-6 covers the cost of audits for water and air pollution. The mean reported cost per audit when no discharge measurements are made is about $150 for air pollution sources and about $300 for water pollution sources. The mean reported costs jump to $1,700 and $950, respectively, when discharge measurements are undertaken. Again, the variation in reported costs per visit is very large, as can be seen from the standard deviations reported in the table. Even so, the differences between audit

Table 2B-5.   Auditing Self-Monitoring Sources: Conduct and Content of
Visits and Use of Results

	All sources	Air sources	Water sources
Conduct of audits			
Responses	68	26	42
Announced	0.16	0.19	0.14
Not announced	0.19	0.19	0.19
Sometimes announced	0.65	0.62	0.67
Sum of frequencies	1.00	1.00	1.00
Content of audits			
Responses	68	25	43
Inspect records	0.97	0.96	0.98
Inspect equipment	0.93	1.00	0.88
Measure discharges[a]	0.90	0.80	0.95
Other[b]	0.12	0.04	0.16
Uses and audit data			
Responses	66	24	42
Identify violations	0.12	0.12	0.12
Identify potential trouble spots	0.11	0.04	0.14
Both	0.77	0.83	0.74
Sum of frequencies	1.00	0.99	1.00

[a]Difference between air and water proportions is significant at 5 percent or better.

[b]Other activities undertaken as part of audits consisted almost entirely of one or another version of a laboratory inspection and check on analytical methods. Note that frequencies do not sum to one because in principle every audit could include every content item.

Table 2B-6.   Auditing Self-Monitoring Sources: Costs of Audit Visits

Type of visit	Air sources	Water sources
Responses	13	20
No discharge measurements (mean)	$155	$301
No discharge measurements (standard deviation)	$129	$324
Discharges measured (mean)	$1,725	$955
Discharges measured (standard deviation)	$1,042	$932

costs with and without discharge measurements are statistically significant at about the 5 percent level for both air and water pollution sources. However, the apparent differences between the costs of air and water audits is only that—apparent. The differences between mean audit costs without measurement ($146) and with measurement ($771) across the media are not statistically significant at the 5 percent level and may well be mere artifacts. This implies that there is no point in speculating about reasons for the differences in discharge measurement costs across media, at least not with a view to "explaining" the numbers in the table. Finally, it should be noted that the figures for costs of air pollution audit visits are lower than those calculated in the text for New Mexico.

One final matter deserves attention before the discussion turns to primary agency monitoring. The questionnaire asked what statistical techniques, if any, were used in interpreting the discharge measurements done during audit visits. Only sixteen positive responses to this question were obtained. (Twelve other respondents checked "don't know," three said "none," and one said "not applicable.") It may very well be, however, that the nonrespondents interpreted "statistical techniques" to include only some relatively elaborate operation such as quality control charting or regression analysis. The aim of the question was much broader, and the few positive responses suggest that the range of techniques in use is correspondingly wide. One respondent, in fact, did mention regression analysis, and another said the agency used "running averages," suggesting operation of a quality-control method. More common was one or another method apparently aimed at taking into account discharge and measurement variability, for example: "range of measurement error," "average of 3 tests," "accuracy and precision," "95 percent confidence interval," "average, maximum and minimum," and "range ±20 percent." An additional four responses referred to EPA rules or significance criteria. Thus, it seems likely that most, if not all, agencies use either a formal technique or a statistically grounded rule of thumb in deciding whether their discharge measurements signal violations. It is the largest disappointment of the survey that the right question was not asked to extract descriptions of these techniques. This would have connected the discussions of statistical techniques and choices of sampling regimes in chapters 5 and 6 more neatly to this description of agency practice.

Primary Monitoring Visits

As has already been noted, self-monitoring requirements turn out to be the rule in state enforcement efforts. Therefore, the questions about sources and situations for which the agency itself had primary monitoring responsibility produced few and not very interesting returns. A total of thirty-five respondents said that for at least some sources they did the only monitoring: twenty-four responses involved air pollution sources and eleven involved water sources. Frequencies of what shall be referred to as primary monitoring visits, or just primary visits, were reported by twenty-two agencies involved in air pollution monitoring and only seven water pollution control agencies. The means and standard deviations of

Table 2B-7.   Primary Monitoring Visits: Frequency

	Source categories			
	Air		Water	
	Large	Small	Large	Small
Respondents	22	17	7	7
Mean frequencies (per year)	1.64	0.75	3.71	3.96
Standard deviations	1.22	0.44	4.15	4.40

Table 2B-8.   Primary Monitoring Visits: Conduct, Content, and Use

	All sources	Air sources	Water sources
Conduct			
Responses	21	15	6
Announced	0.67	0.67	0.67
Not announced	0.10	0.07	0.16
Sometimes announced	0.24	0.27	0.16
Sum of frequencies	1.01	1.01	0.99
Content			
Responses	35	24	11
Inspect equipment[a]	0.91	1.00	0.73
Measure discharges	0.69	0.62	0.82
Use			
Responses	33	23	10
Identify violations	0.24	0.30	0.10
Identify trouble spots	0.18	0.17	0.20
Both	0.58	0.52	0.70
Sum of frequencies	1.00	0.99	1.00

[a]Difference between air and water proportions is significant at 5 percent or better.

the reported frequencies (visits per year) are reported in table 2B-7. As in several earlier cases, most of the apparent differences in the table are not statistically significant. Only the difference between the frequencies for small sources of air and water pollution passes the formal significance test. And even here, the great importance of two states that claim to monitor small sources of water pollution once every month or two and the small sample size suggest that caution should be used in drawing any conclusions about differences between the media.

As far as the content of primary visits goes, the results in table 2B-8 show that equipment inspection dominates, with actual discharge measurement not far behind. Except for measurement of stack emission opacity, all discharge monitoring is done with equipment requiring entry to the source's premises. Other items checked by individual agencies include fuel quality for air pollution sources and stream effects for water pollution sources. One interesting finding of this part of the survey is that only a small fraction of agencies never announce visits in advance. This raises questions about what is being inspected and measured relative to the situation during noninspection periods.

Agency Budgets and Manpower

When some of the early respondents to the survey reported that budget or manpower data were "confidential" or "not available," there seemed reason to fear that nothing would be learned from this section of the questionnaire. In the end, however, at least some data were collected on manpower from fifty-six responses: twenty concerning air monitoring, thirty-four water monitoring, and two involving only manpower totals for both media. On the budget side fifty-two of the returned survey forms contained at least some information: eighteen for the monitoring of water pollution sources, thirty three for the monitoring of air pollution sources, and one with only some total budget data not separable by media.

For the most part the data collected on these two matters reflected the same high level of variability across states already observed in assessing the number of sources covered by the agencies. (For example, while the average complement of "field and lab technicians" over all reporting agencies was 23, the range of complements was from 2 to 290.)[5] Therefore, it will be more instructive to look only at measures of size and budget corrected in various ways for underlying differences in the size of the job to be done. The measures chosen here include the number of annual auditing and monitoring visits per technician, total reported budget per technician, and calculated average costs per visit.

While the number of field and lab technicians and the total agency budget, where they were reported at all, were straightforwardly taken from the survey form, calculating the number of monitoring or auditing visits per year conducted by the responding agency involved several more or less tenuous assumptions. This was because the underlying responses, numbers of sources, and frequency of visits to each source were often hard to interpret or to combine. For example, it was common to find that the agency knew the total number of pollution sources for which it was responsible, claimed not to know how many of these were self-monitored sources, but was confident that there were no sources

[5] The average numbers of air and water pollution lab and field technicians were not significantly different; therefore, only averages across all responses are reported.

for which it supplied the only monitoring. Variations on this theme of apparent inconsistency made it difficult to know how to apply the (auditing and primary monitoring) visit frequency responses. Further problems could and did occur within those responses as well, however. For example, an agency that claimed it had no sources for which it provided the sole monitoring might still answer the question about the frequency of primary monitoring visits. Overall, the result made it seem foolish to estimate a single number for annual visits to sources for each agency. Rather, the maximum and minimum numbers of visits were calculated, the former based on very generous interpretation and combination of frequency and source number answers, the latter on very modest interpretations. The resulting range in calculated visits per year across the responding agencies was enormous—a magnification of the previously noted variation in numbers of sources for which the agencies are responsible. However, it seems likely that for each agency the actual number of visits conducted per year is somewhere between the calculated minimum and maximum. Thus, the ranges reported below for visits per technical person and average cost per visit may also give at least a useful first cut at these interesting and potentially important numbers.

Visits per lab and field technician (calculated from reported technical staffs and ranges of total annual visits—fifty-one responses usable) range from a minimum of zero for several states to a maximum of over 2,200 (water auditing and monitoring visits). The mean of the maximum calculated annual visits was 270 (with $\hat{\sigma} = 417$), and the mean minimum was 56 (with $\hat{\sigma} = 68$).[6] It seems most unlikely that the larger figure is close to the "real" number for it would imply more than one visit per technical person per working day. The lower figure, on the other hand, implies only slightly more than one visit per person per week. It is tempting to guess that this lower figure may be a reasonable clue to actual practice, but this conclusion is necessarily based on only a rough weighing of two counterbalancing observations: visits involving measurement of discharge must certainly involve more than one technical person and probably take the better part of a working week when lab and paper work are counted; but by no means all auditing and monitoring visits include discharge measurement.

Total reported annual budgets per lab and field technician were calculated for sixteen air and for thirty-one water monitoring agencies. The numbers varied from a low of less than $20,000 for one water agency

---

[6] The differences between the means calculated for air and water pollution sources were not statistically significant and the means by media themselves are close to the overall means.

to a high of almost $350,000 for one air agency. It seems nearly certain that this range reflects differences in what is being counted by different agencies rather than simply differences in salary scales and lavishness of equipment. Such reporting differences arise naturally because of the different administrative structures across the states and the different places within those structures of the monitoring and enforcement functions. For example, one important source of variation is the budgetary and administrative treatment of enforcement lawyers—whether they are in the monitoring agency or subagency itself, assigned somewhere else in the environmental agency, or part of a state department of justice. It was not usually possible to sort this out.

The mean total budget per technical or lab employee across all reporting agencies was $44,700, with standard deviation of $21,800. While the means were not significantly different for air and water agencies considered separately, it was true that the variation within the set of air agencies was relatively much less. The ratios of standard deviations to means (the coefficients of variation) were respectively:

For water agencies: $\dfrac{71,400}{76,000} = 0.94$

For air agencies: $\dfrac{20,000}{43,800} = 0.46$

The average cost of a monitoring visit can be calculated from total budget (cost) per technician and visits per technician, since both budgets and visits cover the same time unit, the year. Using the maximum visits figure gives a minimum cost per visit and vice versa. Again, there is considerable variation within the sample and, also again, apparent differences between air and water monitoring operations turn out to be statistically insignificant. Indeed, even the differences between maximum average cost and minimum average cost are not significant for either air or water agencies separately. Looking at air and water monitoring agencies together, however, the following pattern emerges:

	Calculated maximum average cost per visit	Calculated minimum average cost per visit
Mean	$4,275	$844
Standard deviation	$8,117	$1,511

And despite the wide variations, the difference between maximum and minimum is significant at the 5 percent level, because of the increase in sample size.

Perhaps the best that can be done, therefore, is to look at the grand mean—the average of all the calculated per-visit costs, maximum and minimum, for air and water agencies. This is about $2,350 or, in round numbers, $2,500, and may be useful for estimating the long-run costs of various alternative monitoring or auditing plans.

One final note about costs per visit. It might be hypothesized that the responses to questions about the costs of auditing or monitoring visits can be identified with marginal costs. The calculated costs per visit, based on budget totals, clearly are average costs. Because equipment and other overhead items are shared over visits, one would expect to see marginal costs less than average costs. Furthermore, visits during which no monitoring takes place should be the cheapest. If minimum average cost can be roughly identified with average cost of a visit at which no monitoring takes place, the following ranking of available cost numbers might be expected:

Reported cost per visit *without* discharge measurement
less than
Calculated *minimum* average cost per visit
less than
Reported cost per visit *with* discharge measurement
less than
Calculated *maximum* average cost per visit

While the overall means (taking air and water pollution agencies together) do appear to fall in this pattern, none of the differences between adjacent items is significant.[7] Therefore, no evidence is found to support the hypothesis. This can only reinforce the cautionary notes on interpretation sprinkled throughout this appendix.

References

Abbey, David, and Winston Harrington. 1981. "Air Quality Regulation and Management in the Four Corners Study Region: Controlling Stationary Source Emissions." Prepared for the National Commission on Air Quality (Los Alamos Scientific Laboratory, Los Alamos, New Mexico).
Air Pollution Control Association. 1982. "Continuous Emission Monitoring:

[7] The overall means are: reported cost without measurement, $242; calculated minimum average cost, $844; reported cost with measurement, $1,259; calculated maximum average cost, $4,275.

Design, Operation and Experience," *Journal of the Air Pollution Control Association* vol. 32, no. 7 (July) pp. 701–715.

Booz-Allen Hamilton. 1979a. "Evaluation of North Carolina's Program to Regulate Air Pollution from Stationary Sources." Prepared for EPA and CEQ.

———, 1979b. "Evaluation of the City of Houston (Texas) Department of Health's Programs to Regulate Air Pollution from Stationary Sources." Prepared for EPA and CEQ.

Connecticut Enforcement Project. 1975. "Economic Law Enforcement" (Washington, D.C., Environmental Protection Agency) EPA-901/9-76-003.

Environmental Health Letter. 1984. "EPA Seeks to Establish More Consistent Civil Penalty Policy" vol. 23, no. 7 (1 April) p. 2.

Environmental Law Institute. 1979. *The Response to State and Local Regulations on Emissions to the Atmosphere* (Washington, D.C., ELI).

Frye, Russell S., and Karl C. Ayers. 1981. "Air Permits for New and Modified Sources: The Significance of June 8, 1981," *Journal of the Air Pollution Control Association* vol. 31, no. 4 (April) pp. 397–400.

Harrington, Winston. 1981. *The Regulatory Approach to Air Quality Management: A Case Study of New Mexico* (Washington, D.C., Resources for the Future).

Leith, David, Michael J. Ellenbecker, John D. Spengler, Peter Fairchild, and John D. Sarsfield. 1980. "Assessment of Pollutant Monitoring Technologies." Prepared for the Office of Technology Assessment, U.S. Congress.

Meier, Barry. 1985. "Against Heavy Odds, EPA Tries to Convict Polluters and Dumpers," *New York Times*, January 7, p. 1.

Melnick, R. Shep. 1983. *Regulation and the Courts: The Case of the Clean Air Act* (Washington, D.C., Brookings Institution).

Pacific Environmental Services. 1978. "Description of the Regulatory Process for the San Diego Case Study." Prepared for EPA and CEQ.

———. 1979. "Description of the Regulatory Process for the Chicago Case Study." Prepared for EPA and CEQ.

Rossnagel, W. B. 1978. "Comparison of Source Emission Limits" in Paul N. Cheremisinoff and Angelo C. Morressi, eds., *Air Pollution Sampling and Analysis Desk Book* (Ann Arbor, Michigan, Ann Arbor Science Publishers).

U.S. Environmental Protection Agency. 1981. *Profile of Nine State and Local Air Pollution Agencies* (Washington, D.C., U.S. EPA, Office of Planning and Evaluation).

———. 1984. "A Framework for Statute-Specific Approaches to Penalty Assessments: Implementing EPA's Policy on Civil Penalties," EPA General Enforcement Document #GM-22 (16 February).

Uthe, Edward E., John M. Livingston, Clyde L. Witham, and Norman B. Neilsen. 1981. "Development of Measurement Methodology for Evaluating Fugitive Particulate Emissions," EPA-600/2-81-070 (Available from National Technical Information Service, Springfield, Virginia, PB81-196594).

3
Excursions into Law and Technology

The previous chapter of this book assessed the actual monitoring and enforcement practices that are used by state environmental management agencies today. At this point, it seems useful to look at two important elements that influence the strategies chosen by the environmental agencies. The first concerns the law—in particular, how the courts have decided cases involving stochasticity and error as well as restrictions on agency monitoring behavior. The second is the technology of monitoring pollution discharges—its costs and its characteristics, as they relate to both legal restrictions and errors in measurement.

Each of these is worthy of its own full-length review and assessment. And, for specialists, particular facets of the law or particular pieces of monitoring instrumentation alone could probably inspire monographs. This chapter is not a substitute for such efforts. Rather, the major goal here is to provide background and perspective for the more abstract chapters to follow. A secondary goal is to highlight obvious gaps in knowledge as a guide to future research.

The Law and Stochasticity

At an early stage in this study a decision was made to seek court cases involving, as a central issue, the definition of violation in a world of variable discharges and monitoring instrument error. (The technicalities of statistical error and uncertainty will be treated in chapter 5.) The search for such case law proved frustrating, with the available violation cases involving, perhaps understandably, quite different issues. Thus, reports of conviction for violation most frequently seem to involve fraud and falsification of records rather than arguments over the meaning of measurements (for example, Air/Water Pollution Report, 1984; *Virginia Water*, 1983). Some relevant guidance, however, is available. For example, a newsletter for state water pollution control administrators contains a summary outlining how state agencies should judge noncompliance with National Pollution Discharge Elimination System (NPDES) permit terms (Association of State and Interstate Water Pollution Control Administrators, 1983). Several statements in this summary imply a concern with measurement error and "natural" variation in discharges. Some examples are given here.

- Single event violations (that is, of daily maximum limits) and short term violations (that is, of seven-day averages) are discretionary with respect to their designation as significant noncompliance.
- The Director also may consider the significance of violations detected during compliance inspections by using a single event criterion.
- Significant noncompliance for monthly average limitations is based on exceeding Technical Review Criteria (TRC) (magnitude) for a specified time period (duration). The TRC's are for two groups:
  Group I—Inorganic and Oxygen Demand Pollutants (such as BOD, COD, TSS, nutrients)   TRC = 1.4
  Group II—Toxic Pollutants (such as heavy metals, cyanide, and organics)   TRC = 1.2
- The duration is evaluated for any consecutive six months. For all permittees, significant noncompliance is exceedance of the TRC for a monthly average for any two months in a *three* month period.

Thus, the guidelines tell the agency to exercise its discretion in dealing with short-term violations, forgiving some unspecified number of its own choosing. This allows for short-term loss of control on the part of the source that *intends* to comply but also gives more scope for intentional noncompliance. In the longer term, the discretionary element is replaced by a factor (the TRC) by which the permit limit is multiplied to obtain what amounts to a limit on complying discharges. (In chapter 6, this sort of limit will be called an "upper control limit.") The allowance for involuntary loss of control is increased further by the duration specification, that the upper control limit must be breached for two out of three months.

This concern with defining a violation must be matched in a logically consistent way with a definition of the standard itself. And in the area of water and air pollution discharge permit terms, which set the discharge standards for individual sources, there has been a substantial amount of litigation on issues surrounding the statistical extension of the technological bases for these terms. The disputes generally have been over whether and to what extent the Environmental Protection Agency (EPA) was *allowing for* normal variations in discharges, in the sense of making discharge limitations loose enough that violations could not occur when a source was making a good faith effort to comply.[1]

In brief, the process of defining the "effluent limitations guidelines," on the basis of which individual permits are written, depends on ascertaining how particular technologies for discharge reduction actually operate.[2] The data on actual operation are supposed to reveal not only some average reduction possibility but also how discharges vary with changes in the source's operating conditions. The process of standard definition for new sources of air pollution, for example, is described in the following quote (Pahl, 1983, p. 470):

> Numerical emission limits that reflect the performance of Best Demonstrated Technology (BDT) are established for each affected facility. An extensive data base must be acquired to demonstrate that the emission limit is achievable under all the operating conditions that would be considered normal for the affected facility. This is necessary because the emission limit must reflect the performance of the BDT under all normal variations in feedstock, process conditions, production rate, and long term control device performance encountered in affected facilities throughout the entire industry. Thus, the numerical emission limit establishes the minimum emission reduction that would be expected from application of BDT.

Thus, the process of setting emission standards is supposed to rule out the possibility that normal events, including some control equipment breakdowns, will lead to violations. The word "all," taken literally, of course, implies an impossibly loose standard, so that it seems this process also must be understood as aiming at ruling out *all but* some small (but unspecified) fraction of possible events.

While the cases arising from the definition of standards have involved a range of other issues, such as whether EPA used appropriate statistical techniques and whether the appropriate technology was applied or applied correctly, the fundamental question has concerned the extent to

---

[1] In the generally accepted terminology, the sources want generous provisions for "upsets" or "excursions" when emissions are large relative to mean emissions from a plant through no intention of the plant operator. Intentional noncompliance may involve bypassing of treatment equipment.

[2] This discussion ignores requirements for cost estimation, cost-benefit, and regulatory impact analysis.

which EPA should be required to forgive in advance all or some excursions above the standards. Thus, one position argued by plaintiffs has been that if the standard is set so that 99 percent of expected discharges from a conforming plant are below it, then the 1 percent of expected violations should be excused in advance. For example, in the context of a daily standard, four daily violations per year might be excused for a plant operating 365 days per year.

The circuit courts have been divided on this question, with some agreeing that sources attempting to comply should not face inevitable violations because of normal variations in operating conditions. These courts, in effect, have required EPA to remove in advance its discretion to decide how to treat an observed excursion above the standard. Other circuits have held that a successful effort to enforce the pollution control laws must rest on definite numbers defining a violation, although how violations should be treated could remain a matter for agency and prosecutorial discretion. Table 3-1 summarizes a number of relevant court decisions. Of these cases, five were decided in favor of EPA, which is to say for discretion and against explicit advance forgiveness for excursions. Three were decided the other way, with the courts agreeing that some sort of excursion provision, guaranteeing that sources attempting to comply should not be found in violation, should be provided explicitly in advance.

It may be useful to highlight several features of this debate. The first is that three causes of excursions ought to be distinguished a priori. One is normal variation in operating conditions. A second is an abnormal event, such as a breakdown or spill that might reasonably be held to be out of the control of the source. A third is a willful attempt to circumvent the law by, for example, turning off pollution control equipment. These causes cannot be distinguished by measurement: or, said a slightly different way, the responsible monitoring agency cannot be sure on the basis of the measured discharge level that an observed excursion is of the first type, in which case an excuse would be in order, rather than the third type, in which case some enforcement action would seem appropriate. Supplementary information may allow an inference on this matter, but there can be no certainty as to cause. Thus, while an inflexible decision to treat every excursion as a prosecutable violation inevitably will sweep up some intending compliers along with some willful violators, any form of automatic excuse will have the opposite effect, allowing some intentional violators to be free of enforcement action.

This choice between two kinds of error is exactly the choice faced whenever imperfect measurements must be used to decide on the truth or falsity of an empirical hypothesis. The general statistical problem will

Table 3-1.   Summary of Circuit Court Decisions on Defining and Excusing in Advance Violations

Case	Year	Circuit	Medium	Outcome
API v. *EPA*	1976	10th	Water	Refused to require automatic 1 percent of days excursion allowance for refiners. [EPA]
CPC International v. *Train*	1976	8th	Water	Rejected complaint and request for explicit excursion allowances. [EPA]
FMC v. *Train*	1976	4th	Water	No violations should be possible for willing compliers—excursion provisions should ensure this.
Marathon Oil v. *EPA*	1977	9th	Water	EPA required to make upset provisions to cover expected violations.
Weyerhaeuser v. *Costle*	1978	D.C.	Water	Refused to require upset provision (preserving an enforcement system based on straightforward numbers and discretion). [EPA]
Corn Refiners Assoc. v. *Costle*	1979	8th	Water	Refused to require upset provisions (agency should have discretion; alternative is an excessively complicated system). [EPA]
National Lime Assoc. v. *EPA and D.M. Costle*	1980	D.C.	Air	EPA had to take normal variations into account. Rejected idea that discretion was required.
American Petroleum Institute (API) v. *EPA*	1981	5th	Water	Polluter should have to prove an exceedance was reasonable. Also settled a technical question of where requirement should be inserted: in regs or in NPDES guideline documents for industries. [EPA]

Note: [EPA] indicates case decided in favor of agency position.

be discussed in chapter 5. But at this point it suffices to say that the automatic excursion excuse simply changes the balance of errors in favor of the sources.

A second matter that should be noted is that the difference between excusing a certain number of excursions and adjusting the standard to reduce the expected number of unintended excursions can be a significant one. Consider again the example above in which the standard was set so that on 1 percent of the days in a year a source intending to comply might expect to be found in violation because of normal variation in factors underlying its performance. Excusing in advance four days on which excursions were observed would reduce the expected violations to zero. To achieve this effect by adjusting the standard would require, in principle, an infinitely large discharge standard if the results of normal variation are discharges distributed according to the normal or Gaussian density function. Then no source could ever be found in violation. Even using more practical approaches such as, for example, establishing the standard above the highest observed discharge in the underlying source data, the difference between excusing a number of incidents and excusing discharges up to a new limit in general will be different in terms of their impact on average behavior of complying sources and total discharges to the environment.

Notice, finally, that when the actual monitoring and enforcement system contains a combination of formal statistical calculations, such as those reflected in standard setting, informal and ad hoc allowances, such as the "technical review criteria" set out in ASIWPCA, 1983, and prosecutive discretion, the result is to obscure completely the effective probabilities of errors of the two kinds identified above. The direction of change, from formal definition through to discretion, is toward lower probabilities of false violation reports and higher probabilities of failing to identify true violators.

However that may be, it appears that the fundamental issue—whether or not it should be possible, as a matter of law, for a source that is trying to comply to be found in violation—is far from settled. Indeed, the extant cases do not even agree on how the issue should be framed. But perhaps the comment with the most significance for the long run, which comes from the Eighth Circuit in *Corn Refiners Association* v. *Costle* (1979), is that any attempt to excuse in advance any excursion over the permitted discharge limits very well could make enforcement of those limits nearly impossible. At some point, in other words, attention must shift from the intention of the source, which is difficult or impossible to observe, to the actual measured discharges. And some probability that a source trying to comply will be charged with a violation must be accepted. As a practical matter, of course, it is unlikely that a source with very large excursions really is trying to comply, but the point of principle must be accepted before the debate can focus on the real problem—the choice of acceptable error probabilities. Once this is done, it may be possible to make other than an ad hoc choice of acceptable error levels. Eventually a conclusion structured on these lines and on acceptance of the associated errors seems inevitable. Such an approach is set out in chapters 6 and 7.

The Law and Entry for Monitoring Purposes

With perfect discharge control, perfect monitoring instrumentation, and sufficient advance notice of an effort to measure discharges, a polluter need never be found in violation, nor need it ever actually be in compliance except during the actual monitoring operation. Thus, one very practical legal issue is the extent to which advance notice of source monitoring *must be* provided by the responsible agency. The empirical issue, whether or not such notice is actually given, was covered to the extent possible with available data in chapter 2 and its appendix. There, it was reported that most state agencies do provide advance notice of monitoring visits. This presumably reflects uncertainty about the law.

The extent of inspectors' rights of access only recently has been litigated at the circuit court level, although cases involving similar administrative inspections under the Occupational Safety and Health Act (OSHA) and Federal Mine Safety and Health Act (FMSHA) have reached the U.S. Supreme Court. The points at issue have been the necessity of obtaining warrants where consent to enter has not been granted and the type of warrant then required. A related inquiry concerns the amount of notice that must be given to plant operators before each visit.

To date, there has been a consensus among the circuits that a warrant is required under the Clean Air Act. This rule was developed by following the precedent set by the Supreme Court in *Marshall* v. *Barlow's, Inc.* (1978), which held that warrantless inspections under OSHA violated the Fourth Amendment prohibition against unreasonable search and seizure.

The basic framework for search and seizure analysis was laid down by the high court in *United States* v. *Katz* (1967), which concerned government eavesdropping on a telephone conversation. The opinion in this case defined unreasonableness in terms of a person's "reasonable expectation of privacy," through application of a two-pronged test: first, does the person exhibit an actual expectation of privacy; and, second, is the person's subjective expectation of privacy one that society views as objectively reasonable.

The Court applied the *Katz* rule to entry onto private premises in *Camara* v. *Municipal Court* (1967). *Camara* upheld a citizen's right to refuse warrantless nonemergency inspections of his home for possible violations of city housing regulations. This overruled *Frank* v. *Maryland* (1959), which had held that warrantless searches designed to enforce health regulations were not subject to Fourth Amendment prohibitions. In *Camara*'s companion case, *See* v. *Seattle* (1967), the Court applied the *Camara* rule to administrative inspections of commercial premises.

The Supreme Court then proceeded to carve out exceptions to *Camara* for certain industries in which the property owners were deemed to have no reasonable expectation of privacy. In *Colonnade Catering* v. *United States* (1970) and *United States* v. *Biswell* (1972), the liquor distribution and gun sales industries were found to be so "closely" or "pervasively" regulated as to preclude a reasonable expectation of freedom from warrantless entry.

Under this test, the Court in *Barlow's* held that OSHA's statutory authorization for warrantless inspection was unconstitutional. Although the Court specified that its holding was limited to OSHA, this has become the rule for all administrative inspections.

The *Barlow's* rule was further qualified by the Supreme Court in *Donovan* v. *Dewey* (1981), which held that warrantless inspections of mines under the FMSHA are constitutional. In so doing, the Court distinguished the structure of FMSHA's inspection scheme by citing the statute's comprehensive and predictable procedural provisions. Referring back to *Colonnade* and *Biswell,* the Court stated that warrantless inspections are permissible where Congress has determined that such searches are "necessary" to further the federal regulatory scheme and where the regulatory presence is "sufficiently comprehensive and defined" so that the property owner cannot help but be aware that his property will be subject to periodic inspections.

Neither *Barlow's* nor *Dewey* gave a clear explanation as to when a warrant is required by the Fourth Amendment. However, the circuit courts have been unanimous in considering EPA's scheme more similar to OSHA's than FMSHA's, and thus have required warrants.[3]

Remote Monitoring Instruments: A Way Around the Warrant Issue?

If enforcement is in fact hobbled by the necessity, whether absolute or only part of the practical world of the actual regulator, of obtaining each source's permission before entering its plant to measure emissions, can technology solve the problem by allowing remote emission measurement? The answer to this question is far from clear, but it will be useful to distinguish the technological from the legal components of the answer.

The technological answer is that there do exist new instruments capable of producing remote mass emission measurements for several major air pollutants. In addition, plume opacity, which is frequently one of the standards set in permits, has been remotely measurable for a long time. All the relevant devices are optical, using a variety of techniques and subject to a variety of limitations. (For technical de-

[3] On the separate question of using contractors for monitoring, see *AIR/WATER Pollution Report*, 1984b.

scriptions, see Prengle and coauthors, 1973; Hinkley, 1977; Patel, 1978; and Maugh, 1981.) The salient facts, however, appear to be these.

- The general method with the widest applicability to air pollution monitoring involves the measurement of the absorption spectrum of energy to identify and estimate the concentration of various gases in a plume. The source of absorbed energy is often a laser, either of fixed light-emitting characteristics or capable of being tuned to provide a variety of wavelengths. A fixed-wavelength instrument would be one designed to detect and measure a single gas such as SO_2.

- Lasers are available that could be used to detect and measure concentrations of many gases subject to emission regulation, such as NO_2, NO, and CO, but SO_2 is the gas for which most development work has been done.

- Using laser instruments it is also possible to measure plume velocity and, thus, with knowledge of stack exit size, to estimate the mass emission rate for the gas(es) of interest.

- Tests of the performance of these methods as measured against the standard (Reference) methods defined by EPA for extractive sampling of stack gases have been performed and are reported in, for example, Herget and Connor, 1977.[4] Unfortunately, the reports are difficult to interpret within the framework developed in this book, since for the most part the numbers given appear to be for the average errors, without information on the dispersion around the means. For most of the instruments described by Herget and Connor, these "accuracies" are reported to be "about" ±25 percent when concentrations are being measured. For velocity measurements, one of the tested methods (infrared television) gave a concentration value averaging half that determined from in-stack measurements. The other, infrared laser-Doppler, produced values averaging within 14 percent of the in-stack measurements.[5] The remote measurement of opacity apparently can

[4] The methods tested by Herget and Connor were: infrared gas-filter correlation radiometry for SO_2 concentrations; ultraviolet matched-filter correlation spectroscopy for SO_2 concentration; infrared and ultraviolet television for velocity and SO_2 concentration; infrared laser-Doppler velocimetry for plume velocity; and visible laser radar (lidar) for plume opacity.

[5] Herget and Connor comment: "Since Reference Method 2 [the EPA standard for velocity measurement] is considered accurate to within ±20 percent, the agreement [of the laser method with the Reference Method] is about as good as might be expected." This reassurance seems misplaced, though without further information it is difficult to say anything more than that.

be done so that in-stack measurements are reproduced much more faithfully. At a range of 200 yards, the average differences between remote and in-stack readings were on the order of 3 percent for very low opacity levels, and 8 percent at the 30 percent opacity level, though the remote measures were consistently high (that is, biased).

- The costs of this technology are difficult to determine in any fully satisfactory way, since the papers describing it tend either to ignore cost or to be promotional and therefore probably unreliable. Two estimates that do not suffer from the latter affliction are to be found in Maugh, 1981. There, a sophisticated laser instrument called DIAL, designed to measure SO_2 concentrations within a plume, is said to have cost $200,000 to construct on a one-time basis. A simpler instrument, using the sun as its ultraviolet light source but able to measure both concentration and velocity and hence emissions from a stack of known diameter, was displayed at a conference in 1981 with a price tag of $32,700. Even the higher price implies quite a modest cost per monitoring visit. For example, using a 10 percent capital recovery factor and assuming 100 visits per year, the cost per visit would be $200 for the more expensive machine. A full estimate of the costs of monitoring using remote methods, however, would have to include the labor costs of travel, set up, measurement, and analysis. The key difference between this method and the extractive Reference Methods is likely to lie in the set-up, measurement, and analysis components of labor costs.

Despite the advantages of remote monitoring equipment, there remain unanswered legal questions about its place in the arsenal of agencies responsible for monitoring and enforcing compliance with pollution control rules and permits. These questions revolve around the interpretation of Fourth Amendment rights (to freedom from unreasonable searches and seizures) in the context of an openly visible phenomenon such as a smokestack plume.

The Supreme Court held in *Hester* v. *United States* (1924) that "the special protection accorded by the Fourth Amendment is not extended to the open fields." This so-called open-fields exception simply formalizes the common-sense notion that observation of something in plain sight of the public does not constitute a search for purposes of the Constitution.

The exception has never been completely defined by the Supreme

Court, nor has its scope been agreed upon by the circuits.[6] It has been used to uphold diverse activities such as aerial observation through binoculars,[7] *State* v. *Stachler* (Hawaii, 1977); warrantless search of an open barn, *United States* v. *Long* (Eleventh Circuit, 1982); warrantless search of a "junker" automobile in a field, *United States* v. *Ramapuran* (Fourth Circuit, 1980); and warrantless search of a secluded field posted with no trespassing signs, *Commonwealth* v. *Janek* (Pennsylvania Superior).

The open-fields exception was first applied by the Supreme Court to pollution monitoring in *Air Pollution Variance Board of Colorado* v. *Western Alfalfa Corp.* (1974), where it was used to uphold entry onto a plant's premises for the purpose of conducting a Ringelmann test. Western Alfalfa brought the case before the courts, contending that a state health inspector had entered its outdoor premises illegally—without its knowledge or consent—to test the smoke emitted from its chimneys. In a unanimous decision, the Supreme Court stressed that the inspector had not entered the plant itself and had sighted only what anyone in the neighborhood could see. The Court pointed out that while the inspector was on Western Alfalfa's property, there was no evidence that he was in an area from which the public was excluded. "The invasion of privacy . . . if it can be said to exist, is abstract and theoretical."

The uncertain scope of the open-fields exception is demonstrated in the case of *Dow* v. *EPA* (Eastern District, Michigan, 1982), in which the exception was invoked by EPA but found not to apply by a lower court. EPA, while conducting an investigation of powerhouse emissions from Dow's Midland, Michigan, chemical plant, requested permission to enter the plant to take photos of the facility. After Dow denied permission to enter, EPA employed a private company to take aerial photographs of the plant. The photographer used a sophisticated camera to take photos of the unroofed portions of the plant.[8] These photos, when enlarged, showed the equipment of the powerhouse in great detail.[9] EPA asserted that it had two purposes for the overflight: to get

[6] "How Open Are Open Fields?" *United States* v. *Oliver*, 14 University of Toledo Law Review 133 (1982).

[7] "Observation Through Binoculars as Constituting Unreasonable Search," 48 Arizona Law Review 3d 1178.

[8] "Some of the photographs taken . . . at 1,200 feet are capable of enlargement to a scale of 1 inch equals 20 feet or greater without significant loss of detail or resolution," *Dow* at 1357. The Sixth Circuit opinion explicitly dismisses this consideration as irrelevant, however.

[9] Pipes and power lines as small as one-half inch in diameter were visible under magnification.

photographs of the general layout of the plant in relation to the powerhouses and to "confirm excess emissions from powerhouses."

Dow did not learn of the overflight until several weeks later, but then it filed suit in the U.S. District Court for the Eastern District of Michigan to block the use of the photos in enforcement actions.

In the suit, the court was asked to determine:

1. Whether the EPA flyover and aerial photography of Dow's facilities constituted an unreasonable search in violation of the Fourth Amendment;
2. Whether aerial photography of Dow's plant by EPA constituted a taking of property (trade secrets) without due process in violation of the Fifth Amendment; and
3. Whether EPA exceeded its statutory authority under sections 113 and 114 of the Clean Air Act in using warrantless aerial photography as an investigative tool.

Dow sought summary judgment on the Fourth Amendment and statutory issues, and EPA sought summary judgment on all issues. The court declined to rule on the Fifth Amendment claim, citing genuine issues of material fact that made summary judgment inappropriate. The court also declined to rule on Dow's Section 113 question, whether EPA's sole and exclusive remedy when refused entry was to seek injunctive relief under Section 113, asserting that this claim was not ripe for decision as an "actual controversy" had not been presented since EPA had not attempted to secure an ex parte warrant prior to the flyover.[10] The court found that EPA's flight and aerial photography violated Dow's rights under both the Fourth Amendment and the Clean Air Act.

If the warrant requirement found in this initial *Dow* opinion were to stand and be applied, it could logically follow that warrants would be required for the future use of enhanced photography, telescopes, infrared sensing, and anything else involving "visually enhanced surveillance." This could render irrelevant the existence of remote monitoring equipment, no matter how accurate, precise, or cost effective. On appeal, however, the Sixth Circuit Court decided in favor of EPA, rejecting the argument that such remote sensing was unconstitutional (*Environmental Law Reporter*, 1984). Specifically, the court held that the aerial

---

[10] For an ex parte warrant, the person or firm to be subject to search is not represented in the hearing for the warrant. Thus, advance notice is limited to the time of presentation of the warrant.

surveillance searching for Clean Air Act violations did not constitute a Fourth Amendment search and was within EPA's statutory authority. The way may be cleared for increasing use of remote sensing devices. This decision has been appealed by Dow to the Supreme Court, which has accepted it for decision (*New York Times,* 1985).

The Technology of Monitoring

The overall problem of designing monitoring and enforcement policies is defined in part at any one time by the state of the art of existing monitoring (discharge measurement) technologies. The prospects for technical change in these technologies have implications for the range of realistic policy alternatives in the future. As chapter 2 points out, existing officially accepted technology implies costs per monitoring visit in the several-thousand-dollar range. Furthermore, as just discussed above, these Reference technologies require entry into the source's physical plant, thus raising the issues of gaining advance permission or obtaining a search warrant. Finally, the matter of error in deciding whether a violation has occurred is related not only to the range of normal variation in the source's pollution discharges but also to the imprecision inevitably built into the measurement equipment.

Beyond these fundamentals of cost, operating location, and precision, there are other characteristics of measurement technology with implications for pollution control policy design. These might be grouped under a general heading of "ruggedness"—the ability of the equipment to keep performing up to specifications over time. Two particularly important characteristics here are: 1) the ability to stay in calibration (or to continue to measure accurately); and, 2) the ability to continue to operate at all despite the buffets of normal use.

A complete characterization of the policy problem would include the possibilities of choice along each of these dimensions and would reflect the implied tradeoffs. For example, it might be expected that monitoring instrumentation not requiring entry to the plant would display lower precision in measurement, or that, other things equal, higher cost and lower ruggedness might be implied by higher precision.

Such a complete inventory and characterization, however, is beyond the scope of this study.[11] Instead, this section concentrates on a general description of the several phases of the measurement problem—in order

---

[11] It is worth noting also that so far as could be ascertained, EPA itself does not maintain or publish a catalog or review of monitoring instruments and their capabilities.

to make clear that the problem encompasses several separate steps, not just a single, simple action. In addition, some information on equipment cost will be summarized. These comments and illustrations will set the stage for the models in later chapters, in which monitoring cost and precision figure centrally but without any relation being specified between increasing cost and improving precision.

Monitoring Methods in General

In the case of air emissions, Reference Methods for monitoring have been stipulated by the EPA (Code of Federal Regulations, 1977). Subsequently, monitoring requirements were expanded to include the continuous monitoring of certain gases by new and certain existing industrial sources. (Tabler, 1979, contains a summary of the continuous monitoring requirements; EPA, 1979, pp. 2–8, contains a summary of characteristics of existing sources requiring continuous monitoring.) These continuous monitoring systems are defined by performance specifications set out in detail in 40 CFR 60, appendix B. For water, the permitting procedures for the National Pollution Discharge Elimination System (NPDES) eventually may include similarly comprehensive guidelines for monitoring effluents. But as of October 1982 new permitting requirements (Best Available Technology and Best Conventional Technology effluent limitations guidelines) for several important water pollutants (such as Biochemical Oxygen Demand and suspended solids) had not been promulgated by the EPA (Frye, 1982). A good deal of technological work remains to be done in perfecting sensors and other instrumentation before automation is fully incorporated throughout the water cycle (Pitt, 1981, p. 675).

Monitoring any pollutant involves separate phases with choices available at each one. Sample design, sample collection, transport, analysis, and data reduction practices to some extent can be varied for the same pollutant at the same source. These variations in general will result in differences in the cost, precision, and accuracy of the monitoring methods, and these, in turn, will affect the likelihood of accurately identifying a violation of the standard, and therefore the costs of enforcement. Figure 3-1 depicts the phases of a monitoring method, with variable choices summarized for each phase. In table 3-2, examples of actual methods are characterized using the scheme of figure 3-1.

A major objective of the first two phases of the monitoring method, sample design and collection, is to obtain a representative sample of the effluent stream. To accomplish this it is necessary to consider variations over both space and time in the emission composition and volume.

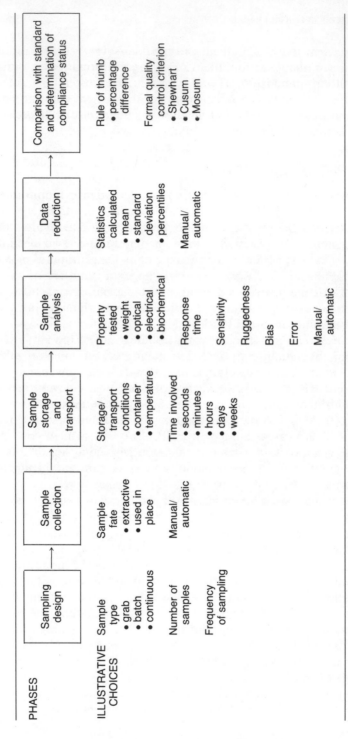

Figure 3-1. Schematic of monitoring process unit activities and choices

78

Table 3-2 Examples of Various Monitoring Methods

Residual	Sample type	Collection method	Storage time	Analytical method	Data manipulations
Particulates (EPA method 5, see below for description)	Grab	Composite, automatic	Hours	Physical (gravimetric), manual	Calculations with other measurement (flow) to get quantity per unit time
Biochemical Oxygen Demand (BOD)	Grab, manual	Composite, manual	Days (difference in dissolved oxygen before and after incubation) 5-day response	Biochemical, manual	Calculations with other measurement (flow) to get quantity per unit time
Opacity (optical in stack)	Continuous	In-situ automatic	Seconds	Optical (transmittance), manual 10 seconds	Averaging over 6 minute period
Metals (atomic absorption spectrophotometer)	Grab	Individual manual	Days	Physical (absorption), manual, seconds	Calculation with other measurement (flow) to get quantity per unit time

Grab sampling is the collection of a single sample at a particular point in the stack or effluent stream at a particular instant. *Batch sampling* is the collection of several grab samples, either: 1) simultaneously from different points in the stream; or, 2) at different times from the same point in the effluent stream. *Continuous sampling* is the continuous collection (and, usually, analysis) of a sample from a single point in the stream. A *composite sample* is defined as a mixture of grab samples collected at the same point but at different times, whereas an *integrated sample* is a mixture of grab samples collected simultaneously from different points (*Standard Methods,* 1976, p. 21).[12] The actual design of a sampling scheme is a complicated matter and will not be discussed here. (For an introduction to the necessary mathematics and techniques, see Box, Hunter, and Hunter, 1978.)

Once a sample has been collected, it is transported to some point for analysis. In the Standard Methods established for water and the EPA Reference Methods established for air, this has meant that the sample (or some extraction prepared from it) is stored for a period of time. In new in-situ gas monitors, samples may be stored for as little as a few seconds or not at all (EPA, 1979). In normal BOD digestion, the water sample can be stored for up to forty-eight hours before analysis (Young, 1981, p. 1,255). For this phase, therefore, it is necessary to decide whether the sample should be stored at all, and, if so, in what form (for example, unmodified, extracted, filtered, refrigerated, heated, preserved, or otherwise) and for how long. Because of the possibility of chemical, biological, or physical changes, long-term storage may render the sample no longer representative of the effluent stream.

After collection, storage, and transport, the sample is analyzed to determine the level of the pollutant of interest. Sample analysis for pollutant levels may be done in the laboratory or in-situ, by extracting a sample from the effluent stream or analyzing it in place, using continuous, automated, or manually operated equipment.[13]

Each analytic method is based on certain physical, biochemical, electrochemical, or other properties of the pollutant. Physical properties

---

[12] This terminological distinction is not universal; in other sciences *composite sample* frequently refers to those collected at the same time from different points.

[13] There is a distinction between continuous and automated analytic methods. *Automated* analysis denotes the use of rapid and automated measurements that are not continuous but, instead, are performed on a grab or composite sample from the effluent stream (Pitt and Denton, 1982, p. 579). This is usually accomplished by replacing manual steps, such as measurement or addition of reagents, with instrumentation. *Continuous* analytic methods depend upon instruments, usually installed on line or in stack, which produce continuous or nearly continuous readings. *Manual* analysis procedures or operation of a piece of equipment involves individual analysis of discrete samples.

measured to determine a pollutant's concentration include weight and optical performance such as absorbance, reflectance, refraction, and diffraction. Biochemical and electrochemical properties are reflected in the chemical reactions a pollutant undergoes, including oxidation, reduction, luminescence, and ionization, and in biological activity, such as toxicity to species of bacteria.

Physical properties, especially size, can be used to separate the desired compound from others for later analysis. Gas chromatography, cyclone samplers, cascade impactors, and filters of all kinds operate under this principle. Gravimetric analyses weigh the sample that has been collected, such as the residue on a filter or that from a chemical reaction. Optical methods are used by much of the sophisticated laboratory analytic equipment, most of the continuous (especially in-situ) monitoring equipment, and the remote instruments discussed in the last section.

Most portable or less-expensive laboratory equipment is based on electrochemical properties. One such device is the ion-selective electrode, which measures the electrical current proportional to the concentration of a given ion that has diffused through a selective membrane. Other electrochemical or electromagnetic properties are used in instruments such as electrochemical transducers, fuel cells, conductimetric and paramagnetic instruments. Chemical and biochemical properties of organic and some inorganic materials are those most frequently used in the Standard Methods for water and the EPA Reference Methods for air. Oxidation-reduction reactions measured by titration or colorimetry are common.

Very often an analysis procedure combines more than one of these principles. The basic idea is first to separate the pollutant of interest and then to measure its concentration. For example, spectrophotometric analysis is frequently performed on a sample that has been extracted by physical separation (such as gas chromatography/mass spectrometry for metals and organics) or chemically (such as most Standard Methods for metals in water) (Vicory and Malina, 1982; *Standard Methods*, 1976).[14]

Because each analytical method has its own idiosyncratic set of interferences (errors caused by the presence of substances not of interest but for one reason or another reflected in the instrument read out), discrepancies will occur when different tests are used for the same pollutant. For example, an in-situ gas monitor may give different results from the EPA Reference Method because entrained condensed water

---

[14] For a terse but comprehensive review of technical literature on analytical methods for measuring chemical species in water, see Shuman, 1980.

Table 3-3. Costs of Analytical Equipment for Measuring Water Pollutants (1983 dollars, rounded to nearest $100)

Pollutant	Analytical method	Analytical equipment cost ($)[a]	Sampling	Site of analysis
Organics	Gas chromatograph/mass spectrometer	17,500 (GC) 70,100 (MS) plus 2,900–6,400 miscellaneous	D	L
Metals	Atomic absorbtion spectrophotometry	14,000–23,400 plus 1,200 miscellaneous	D	L
Metals	Colorimetry	3,500–7,000	D	L
Flow	Turbine, etc.	1,000–2,900	C	I
BOD (DO)	5-day, 20°C	1,300–1,800	D	L
DO	O_2 selective electrode	3,800	[b]	I
Total dissolved solids	Conductivity electrode	400–900	C	I
pH	Glass electrode	200–250	D	L,P
pH	Glass electrode	600–2,500	C	I
Suspended solids	Colorimetry	200–400	D	P
Turbidity	Transmittance	1,100–3,000	C	I
COD	Digestion/reflux/spectrophotometry	7,100 (spectrophotometer) plus 1,200 miscellaneous	D	L

Sources: Alan H. Vicory, Jr., and Joseph F. Malina, Jr., 1982. "Apparatus Needs for Monitoring Priority Pollutants," Journal of the Water Pollution Control Federation vol. 54, no. 2 (February) pp. 125–128; 1982 catalogues and manufacturers' quotes.

KEY: D = grab or batch discontinuous sample; C = continuous sample; L = laboratory; I = in situ; P = portable equipment.

[a] Equipment cost is in 1983 dollars, converted using the Producer Price Index for Environmental Controls, SIC code 3822, from Department of Labor, Bureau of Labor Statistics, Producer Prices and Price Indices for December 1980–1983 (Washington, D.C., Government Printing Office).

[b] Not indicated in source.

vapor in the stack gas acts to scatter light or absorb some of the gases being measured (EPA, 1979).

The final phase of monitoring is data reduction and reporting. In this phase, the data points usually are combined to determine emissions for a given period and are compared with the standard (although opacity, turbidity, color, and similar characteristics do not involve mass emission calculations). Especially when continuous monitors are used, much of the data reduction and reporting can be automated by using systems ranging from strip-chart recorders with manual data compilation to specially designed minicomputers (Maugh, 1983).

Different monitoring methods produce different amounts of data. As stressed above, manual stack or effluent sampling as approved by EPA produces few data points per year because site visits are understandably infrequent. In contrast, continuous monitoring systems for opacity, SO_2, and NO_x register a reading every ten seconds (capacity) or every fifteen minutes (SO_2 and NO_x) (EPA, 1979). Automated monitors for pH, specific conductance, and chemical oxygen demand (COD) are available that can analyze samples at the rate of thirty per hour (Pitt, 1981; Pitt and Denton, 1982).

Costs of Monitoring Equipment and Methods

The cost of a monitoring method includes the purchase of the equipment as well as its operation and maintenance. Fixed costs may include laboratory construction or modification; operating costs must be understood to include all the costs from sample collection through to data reduction. Tables 3-3 and 3-4 list the costs of some selected pieces of analytical equipment used in various monitoring methods for air and water. Each piece of equipment is characterized in the tables first by the pollutant of interest and then by the analytical procedure it employs.

Conclusion

In this chapter, background has been provided for the modeling chapters to follow by explicitly discussing legal and technical problems and possibilities. From these discussions certain conclusions may be drawn, but—as importantly—certain gaps in our knowledge may be identified. The following observations seem especially worthy of note.

The law, as represented by developing case law, has not yet settled on an interpretation of the emission standards and related rules of enforcement being set by EPA. The major unresolved question may be phrased in a variety of ways, but its essence is this: Should it ever be possible for a source that is making a good faith effort to comply with a standard to be found in violation of it because of "normal variations" in its discharges? Asked in this way, the answer seems obvious. If discharges can be very large even while an effort to comply is being made, the answer must be yes or else the standard essentially will be infinity. More difficult and important is what that probability of error "ought" to be. This question lands the courts in the middle of statistical errors— a subject postponed in this book until chapter 5—and forces them to recognize that, unless intent is to become the necessary condition for a finding of violation, innocent sources occasionally will be convicted.

The problem of required notice of intent to monitor is not settled

Table 3-4. Costs of Analytical Equipment for Measuring Air Pollutants (1983 dollars, rounded to nearest $100)

Pollutant	Analytical method	Analytical equipment cost ($)[a]	Sampling	Site of analysis
For sulfur dioxide and other pollutants				
SO_2, NO, NO_2, H_2S, CO_2, O_2	Polarography (electrochemical transducer)	1,700 13,400[b]	D	I
SO_2, NO, CO_2, CO, O_2, opacity	In-situ gas monitor	2,600–8,100 7,600–28,600 33,600–67,200	C	I
SO_2, NO, NO_2, CO_2, O_3	Second derivative spectroscopy	20,700	D	I
SO_2, NO, NO_2, CO_2, CO	Nondispersive infrared spectroscopy (NDIR)	5,100–6,700 4,100–8,700	D	I
SO_2, NO, NO_2, CO	Extractive differential absorption spectrophotometry	20,200–38,600[b]	D	I
SO_2, NO, CO	Extractive differential absorption spectrophotometry	5,100–10,100	D	I
SO_2	Nondispersive infrared spectroscopy	6,500	D	I
SO_2	Fluorescence	10,100–11,800	D	I
SO_2	Flame photometry	5,000–15,100 19,600	D	I
SO_2	Amperometry (coulometry)	8,400	D	I
SO_2	Correlation spectroscopy	9,500–12,500 18,000	D	I
SO_2	EPA method 6 sampling train	8,000–9,500	D	L
For nitrogen oxides				
NO, NO_2	Chemiluminescence	6,700–10,100 9,600	D	I
NO_x	EPA method 7 sampling train	7,500–9,000	D	L

		Cost[a]		
For hydrogen sulfide				
H$_2$S	Polarography (electrochemical transducer)	400	D	P
H$_2$S, mercaptans	Colorimetry	10,200	D	c
For carbon monoxide and dioxide				
CO$_2$, CO	Nondispersive infrared spectroscopy	18,500–22,000[b]	D	I
CO$_2$	Conductimetric	9,500 13,000	D	c
For oxygen				
O$_2$	Electrocatalytic	2,500–9,700	D	I
O$_2$	Paramagnetic instrument	1,700–5,000	D	I
For opacity				
Opacity	In-situ transmissometer	1,300–2,000	C	I
For particulates				
Particulates	EPA method 5 sampling train	5,700–7,000	D	L
Particle size	Cascade impactor	5,600[b]	D	L
Particle size	Cyclone sampler	7,700[b]	D	L
For unseparated pollutants				
Unseparated pollutants	Triple quadrupole mass spectrometry	367,500 35,000	D	I

Sources: *EPA Handbook*, 1979; Wayne R. Ott and David T. Mage. 1981. "Measuring Air Quality Levels Inexpensively at Multiple Locations by Random Sampling," *Journal of the Air Pollution Control Association* vol. 31, no. 4, pp. 355–369; Craig D. Hollowell, Glenn Y. Gee, and Ralph D. McLaughlin. 1973. "Current Instrumentation for Continuous Monitoring for SO$_2$," *Analytical Chemistry* vol. 45, no. 1, pp. 63A–72A; 1982. Catalogues and Manufacturers' Quotes; Anonymous. 1981. "Analysis of Pollutants," *Environmental Science and Technology* vol. 15 no. 7 (July). p. 735; Thomas H. Maugh II. 1983. "Quadropoles Appear in New Instruments," *Science* vol. 220, no. 4593 (8 April) pp. 178–179.

KEY: D = grab or batch discontinuous sample; C = continuous sample; L = laboratory; I = in situ; P = portable equipment.

[a] Equipment cost is in 1983 dollars, converted using the Producer Price Index for Capital Equipment from the Department of Labor, Bureau of Labor Statistics, *Producer Prices and Price Indices Data for December 1973–1983* (Washington, D.C., Government Printing Office). Note that producer price indices for the SIC Classification for Environmental Controls were not calculated prior to 1980. One instrument is required for each pollutant.
[b] Indicates system cost, otherwise cost is instrument cost only.
[c] Place of analysis not indicated in source.

either. Again, the practical question is not whether, but how much. If the agency must present a warrant and negotiate about entry whenever a source is recalcitrant, it seems likely, though by no means certain, that conditions leading to violations could be cleaned up before actual measurements are made. Furthermore, remote monitoring equipment may or may not provide a solution by allowing monitoring without entry. The future worth of this possibility depends on the ultimate fate of the *Dow* case and presumably other tests representing variations on *Dow*'s situation.

The technology of monitoring has not yet been cataloged and characterized in a way that allows for convenient inclusion of real technological parameters in models of monitoring and enforcement system design. One major problem is that information is available for the analytical heart of each method only. A second problem is that relevant characteristics are not consistently and completely reported even for the analytical equipment. Beyond this, it must be recognized that full costs, errors, and ruggedness capabilities relate to the complete method, including sampling techniques, sample transportation, and sample storage. Until these data are available, models such as those developed in later chapters will remain abstract, and agency decisions on instrumentation will tend merely to repeat conventional wisdom or traditional practice.

References

Air/Water Pollution Report. 1984a. "Boise Cascade Fined $100,000 for Tampered Wastewater Tests," January 9, p. 9.
————. 1984b. "Supreme Court Rules EPA Cannot Use Contractors for Air Act Inspections," January 16, p. 12.
American Chemical Society. 1979. *Cleaning Our Environment: A Chemical Perspective* (Washington, D.C., American Chemical Society).
Anonymous. 1981. "Analysis of Pollutants," *Environmental Science and Technology* vol. 15, no. 7 (July).
Association of State and Interstate Water Pollution Control Administrators. 1983. *STATEments* (Washington, D.C., August).
Box, George E.P., William G. Hunter, and J. Stuart Hunter. 1978. *Statistics for Experimenters* (New York, John Wiley & Sons).
Brenchley, David L., C. David Turley, and Raymond F. Yarmac. 1973. *Industrial Source Sampling* (Ann Arbor, Michigan, Ann Arbor Science Publishers).
Cheremisinoff, Paul N., and Harlan M. Perlis, eds. 1981. *Analytical Measurements and Instrumentation for Process and Pollution Control* (Ann Arbor, Michigan, Ann Arbor Science Publishers).

Code of Federal Regulations. 1977. Title 40 Part 60, Appendix A: *Reference Methods,* and Appendix B: *Performance Specifications.*

Dague, Richard E. 1981. "Inhibition of Nitrogenous BOD and Treatment Plant Performance Evaluation," *Journal of the Water Pollution Control Federation* vol. 53, no. 12 (December) pp. 1738–1741.

Environmental Law Reporter. 1984. "Dow Chemical Co. vs. United States *ex rel* Burford," vol. 14, no. 12, pp. 20858–20861.

Foess, Gerald W., and Wayne St. John. 1980. "Industrial Waste Monitoring: A Statistical Approach," American Society of Civil Engineers, *Journal of the Environmental Engineering Division* vol. 106 (EE5) (October).

Frye, Russell S. 1982. "NPDES Rules for Industrial Dischargers: New Obligations, New Opportunities," *Journal of the Water Pollution Control Federation* vol. 54, no. 10 (October) pp. 1349–1354.

Fuerst, C. R. G., E. W. Streib, and M. R. Midgett. 1981. "A Summary of the EPA National Source Performance Audit Program—1979." Prepared for the U.S. Environmental Protection Agency, April (Available from National Technical Information Service, Springfield, Virginia, PB 81-199366).

Goklany, Indur M. 1980. "Emission Inventory Errors for Point Sources and Some Quality Assurance Aspects," *Journal of the Air Pollution Control Association* vol. 30, no. 4 (April) pp. 362–365.

Herget, William F., and William D. Connor. 1977. "Instrumental Sensing of Stationary Source Emissions," *Environmental Science and Technology* vol. 11, no. 10 (October) pp. 962–967.

Hinkley, E. David. 1977. "Air Monitoring with Tunable Lasers," *Environmental Science and Technology* vol. 11, no. 6 (June) pp. 564–567.

Hollowell, Craig D., Glenn Y. Gee, and Ralph D. McLaughlin. 1973. "Current Instrumentation for Continuous Monitoring for SO_2," *Analytical Chemistry* vol. 45, no. 1, pp. 63A–72A.

Hollowell, Craig D., and Ralph D. McLaughlin. 1973. "Instrumentation for Air Pollution Monitoring," *Environmental Science and Technology* vol. 7, no. 11, (November) p. 1011.

Hunt, William F., Jr., and Robert C. Beebe. 1979. "Quality Assurance in Air Pollution Measurements: A Specialty Conference," *Journal of the Air Pollution Control Association* vol. 29, no. 7, pp. 699–703.

Kaplan, David J., and David M. Himmelblau. 1981. "A Cylindrical PbO_2 Diffusion Tube for Separating SO_2 from an Airstream," *Environmental Science and Technology* vol. 15, no. 5 (May) pp. 558–562.

Lettenmaier, Dennis P., and Jeffrey E. Richey. 1979. "Use of First-order Analysis in Estimating Mass Balance Errors and Planning Sampling Activities," in *Theoretical Systems Ecology* (New York, Academic Press).

Loftis, J. C., and R. C. Ward. 1980. "Sampling Frequency Selection for Regulatory Water Quality Monitoring," *Water Resources Bulletin of the American Water Resources Association* vol. 16, no. 3, pp. 501–507.

Mandel, J., 1971. "Repeatability and Reproducibility," *Materials Research and Standards,* American Society for Testing and Materials, vol. 11, no. 8, (August) p. 8.

Maugh, Thomas H. II. 1981. "New Ways to Measure SO_2 Remotely," *Science* vol. 212, no. 4491 (10 April) pp. 152–153.

———. 1983. "Quadrupoles Appear in New Instruments," *Science* vol. 220, no. 4593 (8 April) pp. 178–179.

New York Times. 1985. "Supreme Court Roundup," June 11, p. A-16.

Nriagu, Jerome O. 1979. *Sulfur in the Environment, Part 1: The Atmosphere Cycle* (New York, John Wiley & Sons).

Ott, Wayne R., and David T. Mage. 1981. "Measuring Air Quality Levels Inexpensively at Multiple Stations by Random Sampling," *Journal of the Air Pollution Control Association* vol. 31, no. 4 (April) pp. 365–369.

Pahl, Dale. 1983. "EPA's Program for Establishing Standards of Performance for New Stationary Sources of Air Pollution," *Journal of the Air Pollution Control Association* vol. 33, no. 5, pp. 468–482.

Patel, C. K. N. 1978. "Laser Detection of Pollution," *Science* vol. 202, no. 4364 (13 October) pp. 157–173.

Pedalo, E. F., R. B. Strong, and W. B. Kuykendal. 1981. "Continuous Monitoring for Sulfur Dioxide at a Utility Boiler Equipped with a Limestone Scrubber," *Journal of the Air Pollution Control Association* vol. 31, no. 2 (February) pp. 192–196.

Pitt, W. W., Jr. 1981. "Continuous Monitoring, Automated Analysis, and Sampling Procedures," *Journal of the Water Pollution Control Federation* vol. 53, no. 6 (June) pp. 675–678.

————, and M. S. Denton. 1982. "Continuous Monitoring, Automated Analysis, and Sampling Procedures," *Journal of the Water Pollution Control Federation* vol. 54, no. 6 (June) pp. 576–583.

Prengle, H. W., Jr., Charles A. Morgan, Cheng-Shen Fang, Ling-Kun Huang, Paolo Compani, and William V. Wu. 1973. "Infrared Remote Sensing and Determination of Pollutants in Gas Plumes," *Environmental Science and Technology* vol. 7, no. 5 (May) pp. 417–423.

Rosen, Sherwin. 1974. "Hedonic Prices and Implicit Markets: Product Differentiation in Pure Competition," *Journal of Political Economy* vol. 82, no. 1 (January/February) pp. 34–55.

Russell, Clifford S. 1982. "Pollution Monitoring Survey Summary Report" (Washington, D.C., Resources for the Future).

Samuelson, G. S., and John N. Harman III. 1979. "Chemical Transformation of Nitrogen Oxides While Sampling Combustion Products," *Journal of the Air Pollution Control Association* vol. 27, no. 7 (July) pp. 648–655.

Shen, Thomas T., and William N. Stasiuk. 1975. "Performance Characteristics of Stack Monitoring Instruments for Oxides of Nitrogen," *Journal of the Air Pollution Control Association* vol. 25, no. 1 (January) pp. 44–47.

Shuman, Mark S. 1970. "Nature and Analysis of Chemical Species," *Journal of the Water Pollution Control Federation* vol. 52, no. 6 (June) pp. 1083–1155.

Standard Methods for the Examination of Water and Wastewater, 14th ed. 1976. (Washington, D.C., American Public Health Association).

Tabler, Shirley K. 1979. "Federal Standards of Performance for New Stationary Sources of Air Pollution: A Summary of Regulations," *Journal of the Air Pollution Control Association* vol. 29, no. 8 (August) pp. 803–811.

U.S. Environmental Protection Agency. 1979. *Handbook: Continuous Air Pollution Source Monitoring Systems.* (Available from National Technical Information Service, Springfield, Virginia, EPA 625/6-79 005) (June).

U.S. Geological Survey. 1979. *National Handbook of Recommended Methods for Water Data Acquisition* (Reston, Virginia, Department of the Interior).

Vicory, Alan H., Jr., and Joseph F. Malina, Jr. 1982. "Apparatus Needs for

Monitoring Priority Pollution," *Journal of the Water Pollution Control Federation* vol. 54, no. 2 (February) pp. 125–128.

Virginia Water. 1983. "Filing False Reports Brings Fires," p. 11.

Wang, James C. F., Harvey Patashnick, and George Rupprecht. 1981. "Recent Developments in a Real Time Particulate Mass Monitor for Stack Emission Applications," *Journal of the Air Pollution Control Association* vol. 31, no. 11 (November) pp. 1194–1196.

Young, James C., Gerald N. McDermott, and David Jenkins. 1981. "Alterations in the BOD Procedure for the 15th Edition of Standard Methods for the Examination of Water and Wastewater," *Journal of the Water Pollution Control Federation* vol. 53, no. 7 (July) pp. 1253–1259.

4

Economic Models of Monitoring and Enforcement

One major goal of this study has been to suggest ways to improve the design of monitoring and enforcement programs, especially from an economic perspective. Although monitoring and enforcement problems generally have been assumed away in the environmental policy literature, a few researchers have tackled these problems head on. In this section their work is examined, key themes are identified, and major gaps are noted. This will set the stage at the end of the chapter for a model of the voluntary compliance approach.

Background from the Environmental Policy Literature

A useful way of summarizing and connecting the earlier literature is to look at four major choices of assumptions open to builders of environmental policy models. Two of the choices concern the behavior and capabilities of point sources:

- Will the sources cheat if it is in their self-interest to do so?
- Can the sources control their discharge levels exactly?

The third choice concerns the behavior of the pollution control agency:
- Will the agency monitor for compliance with its standards (or accurate payment of charges)?

A related question, when monitoring is the chosen strategy, is:
- Can the agency monitor without error?

In principle, then, there are twelve possible model types.

The result of using these choices to array five models of pollution control policy implementation along a scale running from naive to sophisticated is to be found in table 4-1. This scale places at the naive end model 1 in which complete and exact control of discharges is combined with an assumed effort to comply.[1] In other words, not only are sources good citizens, but they are very capable citizens as well. Hence, there is no point in agency monitoring activity. As already indicated, this model type contains most of the economic literature on environmental policy alternatives, certainly the bulk of the analytical papers on the characteristics of regulatory, charge, and marketable permit systems. (For reviews, see Bohm and Russell, 1985; and Tietenberg, 1980.) Occasionally authors in this literature bring up matters of enforcement as a basis on which to compare policy alternatives, but the comments are entirely qualitative—and often incorrect or misleading, as discussed briefly in chapter 1. The fundamentally unhelpful nature of models that *assume* good faith compliance and thus make monitoring and enforcement nonproblems was the stimulus for some of the extensions noted in the table.

The intellectual foundations for these extensions had been laid by Becker (1968) and Stigler (1970), writing in very general terms about the economics of crime and punishment. (A later paper in this same tradition is McKean, 1980.) These papers contained several interesting insights that would challenge the researchers of the 1970s and early 1980s. Perhaps most importantly, there was a recognition that the problem of detection error in general would have two facets corresponding to the two types of statistical error. Not only would the enforcement agency have to worry about missing actual violations by applying too loose a test or making too little effort, but as it tightened up it would

---

[1] The seven possible model types not represented in the table include two of the four models in which the agency is assumed not to monitor. Both of these would include a source with stochastic discharges; in one case cheating would be possible and in the other it would be assumed away. Two other model types not yet explored would involve pointless monitoring of complying sources with complete control over their discharges. The other three unrepresented models represent variations in the source of uncertainty. For example, the fourth model type in the table involves monitoring *with* error of sources attempting to comply but lacking full discharge control. An alternative model would involve monitoring such sources *without* error. Such differences are unlikely to produce different conclusions, so it seems reasonable to think that the models that have been explored include the important general situations.

Table 4-1. Characterizing Models of Monitoring and Enforcement

Model	Assumptions about firm	Assumptions about agency	Assumed or taken as given	References	Focus of analysis	Other features
1. Common environmental economics (assumed compliance)	Capable of perfect discharge control; complies regardless of self-interest	Does not monitor	Discharge standard or emission charge; firm's costs; compliance	Bohm and Russell, 1985; Tietenberg, 1980	Optimal standards or charges; choice between standards and charges	
2. Product quality standard (some compliance privately optimal)	Capable of perfect discharge control; may or may not comply according to self interest	Does not monitor	Quality standard; firm's costs (reflecting private interest)	Viscusi and Zeckhauser, 1979	Implication of noncompliance for effectiveness of standard	Industry structure instead of representative firm
3. Pollution standard (simplest monitoring case)	Capable of perfect discharge control; may or may not comply according to self-interest	Monitors without error	Discharge standard or emission charge; firm's costs; probability of "visit"; fine for violation	Viscusi and Zeckhauser, 1979; Storey and McCabe, 1980; Harford, 1978	Implications of noncompliance for effectiveness of standard (Viscusi and Zeckhauser); implications of noncompliance for choice between standard and charge (Storey and McCabe, Harford)	Use of control cost subsidies (Harford); effect on output (Harford); Industry structure (Viscusi and Zeckhauser)

4. Pollution standard (stochasticity in measurement)	Stochastic (or) discharges; tries to comply regardless of self-interest	Monitors with measurement error	Discharge target or standard; firm or agency costs; probability density functions for discharges (or measurement errors); benefits of control/costs of false alarm; attempt to comply	Downing and Watson, 1973–1975; Vaughan and Russell, 1983	Implications of stochasticity for optimal choice of technology by firm (Downing and Watson); implications of stochasticity for design of agency monitoring scheme (Vaughan and Russell)	Stochastic control cost functions (Downing and Watson)
5. Pollution standard (stochasticity in measurement and cheating by source)	Stochastic (or) discharges; may or may not comply according to self-interest	Monitors with measurement error	Discharge target or standard; firm or agency costs; probability density functions for discharges (or measurement errors); benefits of control/costs of false alarm	Linder and McBride, 1984	Implications of noncompliance and stochasticity for choice between standard and charge	Decentralized enforcement

confront the other facet—the false alarm or groundless charge of violation, creating costs for society at large and political problems for itself.

The significance of this insight becomes apparent only when there is some element of stochasticity in the system, whether in the detection capabilities of the agency or the discharge-control capabilities of the sources; but the characterization of model 2 in table 4-1 is the rejection of the perfect compliance assumption without the introduction of true stochasticity. In particular, Viscusi and Zeckhauser (1979) analyze a situation in which firms have a private incentive to provide at least some portion of what the regulatory agency wants. In their case this is product quality (measured in a single dimension). In this context it makes sense to examine optimal standard setting when noncompliance is allowed as a possibility but no monitoring effort at all is hypothesized. The lesson of this section of their paper is that in the industry setting, and in the absence of perfectly enforced compliance, raising the product quality standard cannot be assumed to raise average output quality. This is because, as the standard is raised, some firms will drop out of compliance rather than pay the increased costs of higher quality. These dropouts must be balanced against the higher quality produced by complying firms.

When interest is confined to pollution discharge regulation, the possibility that some nonzero control level will be privately optimal seems less likely. Then, monitoring activity and enforcement incentives take center stage and the resulting model 3 in table 4-1 is more sophisticated. Viscusi and Zeckhauser in another part of their paper show how a positive probability of detection of noncompliance affects their results—which is hardly at all except to raise the average level of quality above the no-monitoring case while leaving it still below the level specified by the regulation.

It is interesting, however, to analyze a bit more closely the meaning of this probabilistic enforcement. Because the probability of detection does not vary with the firms' decisions, it can best be thought of as a random decision process run by the agency for determining whether or not to visit a firm subject to the regulation. Once the agency decides to visit, detection of a violation or determination of compliance follows with perfect reliability by assumption. If all firms are visited each period, the probability measure equals one and no violations will go undetected. But neither will any false accusations of violations be made, because there is no true stochasticity in the actual discharges or in the measurement technology.

The model in the pollution control field most closely corresponding to that of Viscusi and Zeckhauser is one published by Storey and McCabe (1980), although here the interest is not in total discharges from an industry. Rather, they deal with the differential implications of imperfect enforcement for alternative implementation systems: the discharge standard and the emission charge. Acknowledging a debt to the earlier work of Allingham and Sandmo (1972), they show how the choices open to the enforcement agency—of probability of detecting a violation and the fine imposed for a detected violation—affect the representative firm's decision.[2] It is significant, however, that the detection probability is not taken to depend on the actions of the firm, which is to say that it is independent of the size of violation of the standard. This assumption is consistent with the random-visit model referred to earlier.

Another model that focuses on the firm's decision problem in the face of uncertain enforcement is that of Harford (1978). This model is classified here as reflecting perfect control of discharges (and perfect measurement by the agency) because that is the spirit in which it is written. But Harford does introduce a detection probability that varies with the size of the violation (exceedance of standard or underreporting for charge payments). This feature hints at the problems to be faced in the next categories of models, those recognizing stochasticity in the deeper sense. This is true because in that more complicated situation, other things equal, the agency will have a higher probability of detecting a violation the larger the violation. On the other hand, Harford does not discuss the agency's problem in this more complex situation, and his paper equally well might be taken to reflect a random visit model complicated by some agency procedure making use of indirect evidence on violation size to influence the probability of a visit.

The only other difference between the Harford and the Storey and McCabe papers is that the latter transforms the firm's profits through a utility function, which allows them to explore the role of risk aversion, while Harford confines himself to expected monetary value analysis. In both papers interest centers on comparative statics, in particular on the effects of choice of policy alternatives on firm decision variables. Both derive results showing that for a discharge-standard approach to pollution control, actual emissions decrease with increases in probability of violation detection and in the size of fines levied for detected violations. Harford finds some ambiguity in the effect of the standard itself, with the sign of the relation between actual discharges and the level of standard depending on the slope of the marginal expected penalty (*MEP*) function. If the marginal expected penalty increases with the size of the

---

[2] Viscusi and Zeckhauser highlight an industry consisting of differentially situated firms. Storey and McCabe are content to look at representative firms. It seems clear that analogs of the effects observed by Viscusi and Zeckhauser at the industry level would crop up were the latter models to be extended in that direction.

violation, discharges will be reduced as the standard is tightened. If the marginal expected penalty is a constant with increasing violation size, tightening the standard will have no effect on discharges. And if the marginal expected penalty falls as violation size increases, a tighter standard will lead to larger discharges. These three alternatives are shown graphically in figure 4-1, where the alternative standards are labelled S_i and the resulting discharges as E_j.

If expected penalty $P(v)$ equals probability of detection $p(v)$ times fine per violation $F(v)$ and if both probability and fine are functions of violation size in general, then:

$$P(v) = p(v)F(v), \tag{1}$$

$$\frac{dP}{dv} \equiv MEP = p'F + F'p \tag{2}$$

and

$$\frac{d^2P}{dv^2} = p''F + F''p + 2p'F' \tag{3}$$

where primes, as usual, indicate derivatives.

Thus, for example, if the probability of detection $p(v)$ is in fact a constant and $F(v) = \beta v$ $(\beta > 0)$ then:

$$\frac{d^2P}{dv^2} = 0, \text{ or case (b) in figure 4-1}$$

If $p' > 0$ and $F = \beta v$ then even if $p'' = 0$
$$\frac{d^2P}{dv^2} = 2p'\beta > 0, \text{ or case (a) in figure 4-1}$$

If $F = \beta v^{1/2}$ and $p(v)$ is a constant
$$\frac{d^2P}{dv^2} = -\frac{1}{4} p\beta v^{-3/2} < 0, \text{ or case (c) in figure 4-1}$$

It may seem puzzling that Storey and McCabe, who have created a constant marginal expected penalty by having a detection probability independent of violation size and a constant fine per unit of violation, can report an unambiguous result for the effect of the standard on actual discharges. The trick is their utility function approach, which effectively

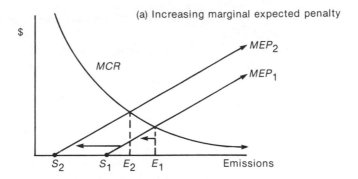

(a) Increasing marginal expected penalty

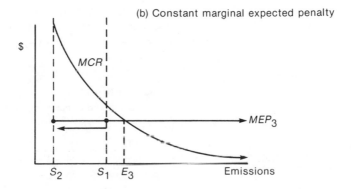

(b) Constant marginal expected penalty

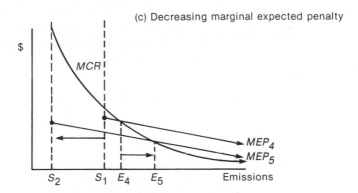

(c) Decreasing marginal expected penalty

MCR = Marginal cost of emission reduction
MEP = Marginal expected penalty

Figure 4-1. Relation between shape of marginal expected penalty function and effect of changing emission standards

translates a constant expected marginal penalty into an increasing expected marginal loss of utility.[3]

Harford is also interested in the role of a subsidy for abatement costs
and in the response of the firm's output to the policy instruments. He
finds that an increase in the subsidy, defined in his terms, will decrease
discharges. And output responses are found to parallel those found for
discharges, except that the response of output to subsidy level is am-
biguous.

Storey and McCabe are also interested in relative effectiveness of
policy instruments in inducing changes in discharge. They recognize that
desirability cannot be judged solely on the basis of "effectiveness" because costs of implementation in general will differ. But even in terms
of effectiveness alone their analysis, confined as it is to the sizes of
partial derivatives, is less helpful than it might be. Had elasticities been
used the comparisons would have been more easily interpreted.

Finally, both these papers also deal with an emission-tax policy alternative, exploring how enforcement choices influence behavior in that
context. The situation modeled in both cases represents a self-reporting
billing system; that is, the emission charge is paid on reported discharges,
which may be different from actual discharges. If the agency detects
such a difference, a violation is taken to have occurred and both a special
penalty and the requirement that the charge itself be paid on total
discharges are imposed.[4]

By far the most interesting finding, one might even call it startling at
an intuitive level, of both the Harford and the Storey and McCabe
analyses of a charge situation is that actual discharges are determined
entirely by the level of the charge, independent of detection probability
and fine structure. The monitoring and enforcement apparatus affects
the firm's choice of *reported* discharges and hence the size of the po-

---

[3] Storey and McCabe set out to help the reader understand their algebraic results by a
parallel graphical presentation. But because the graphs show risk-neutral firms, the authors
are forced to use increasing marginal expected penalty functions to show what is hap-
pening. The discussion in the text (p. 36) in combination with the figures (p. 35) fails to
alert the reader to this point.

[4] Notice that the logic of the models is strained here because of the "timeless" structure
of the comparative statics. To check on the reported discharges the agency must have
been monitoring *during* the period covered by the report. But in these models, if monitoring was going on, it certainly would have detected the true discharges. Thus, there
would be no point in the source underreporting discharge unless it were unaware of the
monitoring; that is, the source would know whether or not to report accurately before it
had to choose. Only if the interpretation of the probability measure is shifted to something
involving the firm's knowledge of the agency actions can one be comfortable with the
approach.

tentially discovered violation. This result depends, however, on there being a nonzero probability of detecting violation and thus a nonzero probability that payment of the correct charge (plus fine) will be required. In the absence of any detection effort the optimal policy for the firm is effectively indeterminate. Although reporting zero discharge and practicing zero control is formally optimal (zero cost), one can observe, as does Harford, that a zero discharge report might be taken as prima facie evidence of cheating, even in the absence of monitoring (detection probability equals zero). Therefore, even though in these models there are no *marginal* effects on actual discharges of increases in detection probability or effective fine for violation, there is a nonmarginal connection requiring a credible effort by the agency.

The other side of the interesting result just discussed is that while the marginal effects of detection probability and fine for violation (underreporting) on *reported* discharges are as would be expected, increases in the emission charge lead to decreased levels of reported emissions. Indeed, Harford shows that under the most reasonable assumptions, this decline in reported discharges will exceed the decline in actual discharges so that the violation level will tend to increase. Thus, increasing the emission charge increases the agency's monitoring and enforcement problem even though actual discharges may move as desired. Harford also shows how this result plays out in the context of the choice of optimal charge levels when damage functions are known, the extra enforcement cost being additive with abatement expenditures, implying that a higher level of marginal damage will be optimal than would be true in the naive model.

Table 4-1 then introduces what might be called "true stochasticity" of discharges or measurement error in model 4; that is, real uncertainty for the agency about whether or not the discharger is violating the standard (or reporting his true discharge level). The easiest way to see how such uncertainty can arise is to think of the agency as possessing a monitoring technology such that the measured value is an unbiased but imperfect estimate of the actual discharge. Then a measured discharge exceeding the announced standard has some probability of being incorrect; the source could be in compliance but still be accused of a violation. Or, on the contrary, a measurement indicating compliance could be consistent with an actual violation. An analogous but more complicated problem exists if the source can only choose the parameters of a probability density function from which its actual discharges will be "drawn" by virtue of random events, such as weather and human error, outside its control. Then a measured value of discharge could be assumed to match the true value perfectly, but there would still be linked questions about how to interpret and announce a standard and how to determine whether a particular measurement indicates a violation.

In the simpler version of the world with stochasticity, the source does not try to cheat. Rather, good faith efforts at compliance are assumed as in the approach characteristic of much of the environmental policy literature. The problem for agency and source is to take account of uncertainty about compliance status, devising respectively optimal monitoring and optimal responses to monitoring in the light of that uncertainty. The second of these problems is analyzed, using a simulation model, by Downing and Watson (1973, 1974, 1975; Watson and Downing, 1976). Their setting is the initial compliance test required under performance standards for new sources of air pollution. In that setting, cheating in the sense of avoiding or behaving differently for monitoring is not really a possibility. The plant (an electric utility boiler is what they have in mind) must be run at capacity for the test, according to the regulation. The test is designed to provide evidence of the *capability* of the control equipment (electrostatic precipitator for the control of particulates in the Downing/Watson work) to meet the emission standard at capacity operation. Passing the test is a necessary condition for actual operation, and the test may be repeated until it is passed. In their model a pass results when the average of three simulated measurements of particulate emissions is lower than the required standard. The emissions in the model are drawn from density functions parameterized to mimic different precipitator options. The firm chooses a precipitator technology based on expected costs of passing the initial compliance tests and the costs of subsequent operation and fines for violations.

It is worth noting that Downing and Watson mean for their work to be more general than the characterization given here. In particular, they preface the simulation model by a "general enforcement model" that includes the key possibility of cheating by the source to achieve apparent *continuous* compliance after startup. Their contribution is discussed here within the no-cheating category because the general enforcement model is not completely or convincingly developed. It is left at the stage of very general functional notation and accompanying text, with no effort to connect textual commentary with complete first order conditions, let alone second order conditions or comparative static expressions for the effect of policy variables. Further, as already noted, the situation sim- ulated with uncertainty built in, initial compliance testing, offers no real opportunities for cheating, while the ongoing compliance problem is modeled in such a way as to eliminate uncertainty. For the latter, a deterministic decline in precipitator efficiency is combined with certain and perfect (if lagged) determination of actual discharges and hence certain detection of violations. Again, cheating is assumed away. Thus, although Downing and Watson made a substantial contribution at a very early date, their discussion of the most complicated situation did not advance matters much beyond the largely literary efforts of the Becker/ Stigler tradition.

In a 1983 paper, Vaughan and Russell took up a different viewpoint and a different problem but also confined themselves largely to the case of monitoring in a no-cheating situation. Because some of the work reported in that paper will be repeated below, it is merely noted here that their paper deals with the agency's design of an optimal monitoring program for continuing compliance. Sources monitored were assumed to be trying to comply but were subject to periodic errors and malfunctions leading to violations.[5] Both random discharges and measurement error were allowed for. To characterize the agency's problem it was necessary to assume that the marginal benefits of controlling each source were known to the agency along with the costs of monitoring. Only when the violations being looked for could be assumed to be of known size was it possible to provide a plausible sketch of a solution method.

These two studies are not alone in their interest in the problems discharge stochasticity poses for monitoring and enforcement, nor in their assumption of good faith compliance efforts on the part of sources. For example, Berthouex and Hunter (1975) and Berthouex, Hunter, and Pallesen (1978) worked on wastewater treatment plant monitoring using a quality control approach. Witten and coauthors (1982) investigated the links between recognized discharge stochasticity, an upper limit standard reflecting that recognition, and the mean discharge target adopted by the source. They also were concerned with the effects of varying the averaging period for standards and of taking discharge autocorrelation into account. Casey, Nemetz, and Uyeno (1983) derive rules of thumb with desirable properties around which to design monitoring programs, where the mechanism producing violations is not specified, except as a random process, but clearly does not involve volition (cheating).

Related work involving the relation between stochastic discharges and attainment of *ambient* standards includes Beavis and Walker (1983). Curran and Steigerwald (1982) write about ambient data, but their demonstration of how measurement imprecision creates bias in estimates of

---

[5] Alternatively one could say that extremely inept cheaters were assumed. These sources follow an exponential decay rule in deciding when to cheat after the last violation has been detected and corrected.

the extremes (for example, the top decile) of distributions is equally applicable to point source monitoring where limits on extremes are involved.

Then model 5 in table 4-1 brings the assumption that cheating is possible into the same problem with measurement or control uncertainty. This is attempted by Linder and McBride (1984) in a stimulating but ultimately incomplete paper. They begin with the agency's monitoring problem in the situation of measurement uncertainty and show how the agency can choose an optimal pair of error probabilities. This choice is based on information about the benefits of discharge control, the costs of false alarms (and the mirror images of these two quantities), plus an a priori judgment about the likelihood that a violation will be attempted and an assumed separation of mean discharges in the violation and no-violation case.[6] In this formulation, past cheating raises the present a priori probability that a violation is being attempted, but cheating plays no other, more essential, role in the agency's decision process. There is no time dimension to the problem so there is no opportunity for the source to bring its discharges into compliance when monitoring is to be carried out, intentionally allowing violations to occur at other times. Nor is the source's action assumed to influence the measurement problem or the costs and damages (except for political pressure that, qualitatively, is said to raise the costs to the agency of false alarms).

The second major section of the paper introduces a modified version of Harford's source problem—the maximization of source's profits subject to a discharge constraint that can be violated. The modification consists of allowing the source some influence, at a price, over the detection probability actually achieved by the monitoring agency. Thus, the source can choose the size of its violation (the size of its discharge) and the extent of its "concealment activity" as well as its production level. Cheating therefore subsumes the following three decisions: to violate the discharge constraint, to determine the size of the violation, and to make efforts to conceal the violation from the agency. The first order conditions derived for this problem are qualitatively the same as those reported by Harford and by Storey and McCabe with modest complications introduced by concealment possibilities. This is also true when Linder and McBride turn to the case of an emission charge payable on the basis of self-reported discharges. Their conclusions about directions of effects, however, are less precise than those of the earlier authors

---

[6] This separation may be thought of in standard deviation units so that the same difference in absolute terms becomes larger, and thus the errors smaller, if improvements in measurement technology decrease the standard deviations of the density functions about their means.

because they did not work out the comparative statics as carefully and depend on verbal analysis of the first order conditions.

Notice the timeless nature of the cheating reflected in this model. The authors are not talking about sequences of events occurring in time— for example, firm decides to cheat by turning off its control device; the agency announces a monitoring visit; firm turns on device; monitoring visit occurs; firm passes; firm turns off control device; etc. Rather, the firm and agency arrive at an expectational equilibrium in a game with mixed strategies, the firm using some random device to decide each day whether or not to turn off its control device, the agency using a similar device to decide whether or not to pay a monitoring visit, and firm having some chance of preventing detection of a violation even if the agency decides to monitor on a day when the control device has been turned off (perhaps by turning it on when the agency vehicle reaches the plant gate). The optimization problem refers to the average day in this ongoing process.[7] (The idea of expectational equilibrium in a monitoring game will be reintroduced in chapter 7).

Other analytical studies relevant to monitoring and enforcement in a world where both stochasticity and the possibility of cheating are recognized are scarce. Greer, Douney, and Tallent (1978) attempted to identify significant factors affecting the decision whether or not to comply with discharge standards. They went beyond the usual economic assumption that cost minimization in the narrow sense was a sufficient motivation by focusing not on the nearly mythical owner-operator but on the manager-operator, who is part of a large bureaucratic corporation. Unfortunately, the study is of limited interest not only because of its imprecise formulation, but also because, in attempting to test their ideas, the authors chose to use a sample of municipal treatment plant operators, a special population where incentive structure is concerned.

In a paper on alternatives to the sort of monitoring and enforcement being discussed here, Wittman (1977) discusses how liability rules concerning damages can usefully substitute for ex ante attempts to control performance. For example, assume it is technically difficult, constitutionally obnoxious, or just plain expensive to monitor some kind of behavior carrying the possibility of causing damage to another person or another's property. Then it will be difficult or impossible to control the behavior and prevent the damage by regulation. On the other hand, in some such cases it may be easier to monitor the actual damaging

---

[7] Alternative interpretations could be worked out for continuous monitoring situations, where imperfect instrumentation and attempts to tamper with monitoring output define the detection probability density function over violation size and the adjustment to the function resulting from concealment activity.

events and assess blame for their occurrence. In these cases, a liability rule provides decentralized incentives for avoiding the potentially damaging behavior and reduces the necessity for regulation with attendant monitoring and enforcement. Traffic laws designed to prevent accidents, monitored by police and reinforced by liability (whether for your own insurance rates or direct cost payments), are used as examples by Wittman. It seems that in the pollution-control field the scope for liability to substitute for monitoring is limited to situations in which cause and effect are clear—that is, where source and damage can be connected. Such might be the case where an industrial plant maintains a waste lagoon on its own property, or where a municipal wastewater treatment plant is the only source on a lake or river. But liability for damage from ambient NO_x or SO_2 pollution in a metropolitan area could be assigned only with grave difficulty and probably never could be used to penalize a specific violation of a discharge standard at a particular plant.

In the course of the above review, several themes have recurred. These include:

- the choice open to the source of attempting or not attempting to comply with the regulations issued by the agency
- the complications for determining compliance status created by stochasticity of discharges (in combination with errors in measurement)
- the choice to be made by the agency of how often to pay a monitoring visit
- the choice of enforcement mechanisms; that is, the choice of punishments for violation of standards or false reporting of discharges
- the characteristics of available (and desirable) monitoring technology, including whether or not it operates continuously, whether or not it requires access to the premises of the source, how precisely it measures discharges (how large its standard error of measurement is), and its cost
- the legal constraints on the agency's freedom of action, such as requirements that prior notification of a monitoring visit be provided to the source.

The challenge for environmental policy analysis generally and for this study in particular is to weave these themes together in a way providing some realistic and useful guidance for state and federal agencies. As a first step in that direction, the next section presents a model exploring the implications of a voluntary compliance approach—that is, no penalties are assessed, but sources attempt, with imperfect success, to comply. No uncertainty is admitted.

Model of a Voluntary Compliance System

In this section a model is explored in which sources are assumed to be trying to comply. Discharges are stochastic, in the sense that a random "failure" of the source's control equipment may occur. But the source does not choose when a failure will occur—in that sense it does not choose to violate. The fact that failures do occur, however, is evidence that the source is conscious that it will occasionally violate the regulation and cannot or will not act to prevent this. It is thus a willing but imperfect complier.[8] Monitoring visits are scheduled randomly by the enforcement agency, and if a visit takes place after a failure has occurred the violation will be discovered with certainty. No fine is levied for a discovered violation, but the source is assumed to act to correct it.

The model does *not* endogenously explain why the sources choose to comply. Rather, it is designed to explore the implications of assumptions that include the no-purposeful-violation assumption. It is shown that if sources make some, but imperfect, efforts to comply, and if, when detected in violation, they voluntarily take steps to return to compliance, the pattern of violation frequencies generated can be made to approximate that observed in real situations. What remains to be provided is some exogenous justification for the assumption that the sources will make efforts to comply.

Continuous Compliance: A Bargain

An explanation of compliance, which is consistent with the rational self-interest assumptions of neoclassical economics and does not require the use of fines or other legal sanctions for each violation incident, does require a broader view of the costs and benefits to sources and agency of various courses of action. This view sees voluntary compliance as a bargain struck between the plant and the pollution control agency. On one side of this bargain the agency agrees to take a tolerant attitude

---

[8] Since the frequency of breakdowns might be thought of as under the control of the source, and a measure of the seriousness of its commitment to compliance, it is perhaps more accurate to say that the decision process that produces the breakdown frequency is simply taken as given.

toward failures in continuous compliance, giving violators a "free" chance to return to compliance before seeking penalties. For its part, the plant agrees to make a "good faith" effort to comply with pollution control regulations. While the precise definition of "good faith" depends on the source and the agency, voluntary compliance clearly requires some degree of observance of emission regulations.

The major reason the system can work is that the costs to both parties of breaking the bargain are substantial. On the agency side, the assurance that sources are making some effort to comply allows for lower levels of surveillance effort. The engineering inspection can be the primary method of surveillance because there is generally no need to gather evidence of a violation that will stand up in court. This takes pressure off the agency's budget and also may allow the inspector to function as a pollution-control consultant, offering advice on how to improve performance. Although the voluntary compliance approach is not necessary for such interaction, it certainly facilitates it. On the other hand, sources considered in "marginal" compliance may be warned to expect more frequent surveillance.

From the source's point of view, there are penalties associated with the loss of voluntary compliance status and these penalties can be considerable. Even ignoring the possibility of an eventual fine, most state agencies also have power to seek an injunction to shut down the offending plant. The costs of a shutdown can be very high. A firm in this situation will also incur legal fees, which for a small firm can be burdensome. Furthermore, recalcitrant firms can also suffer from bad publicity, something likely to be especially important to firms serving local markets. Finally, some agencies have another weapon, at least in principle, which is their control over renewable operating permits.[9]

Notice that these incentives are conceptually similar to the fines that appear in the economic models of enforcement discussed earlier, except that in those models the sanctions are invoked whenever a violation is discovered. In the voluntary compliance approach, sanctions arise only when the agency considers the firm to be uncooperative. Many violations can be committed without the firm's losing its voluntary compliance status. Another difference is that penalties, when they are levied, seem to be related to the degree of recalcitrance rather than the magnitude of the violation.

---

[9] In none of the states surveyed in chapter 2, however, were permit renewals made contingent on past performance; that is, withholding of permits seems to be little used as a means of enforcing compliance.

This argument from the side of prospective penalties is reinforced from the cost side if one takes seriously the contention that much pollution control equipment is capital intensive. Then the prospective savings in variable operating costs from intentional violation may generally be small relative to total annualized costs.

The problem, then, is to define behavior that will not result in loss of voluntary compliance status though neither will it in general produce a perfect compliance record. For example, an operator would be unlikely purposely to bypass abatement equipment unless he were certain that it would not be discovered. But such certainty is very difficult to attain, given the possibilities of surprise inspections or disgruntled employees.[10] Less dramatically, a plant's commitment to voluntary compliance might be questioned if the operator refused to entertain suggestions from inspectors that would improve performance.

Thus, if a firm wished to avoid jeopardizing its voluntary compliance status, it would operate the abatement equipment at its plants in a manner that some fraction of the time resulted in compliance. Plants wishing to reduce the cost of maintaining the voluntary compliance bargain must find more subtle ways to economize than simply turning off abatement equipment. The key possibility is to skimp on maintenance. It is likely to be difficult for an inspector to tell, on a periodic visit, whether maintenance is adequate, and it is therefore difficult to document an assertion that the plant has been negligent. The effect of too little maintenance on equipment performance over time is difficult to generalize about, beyond asserting that performance will deteriorate over time. (This is very similar to part of the Downing and Watson, 1974, argument.)

The average emission rate of a source, then, will be somewhere between the design emission rate of the abatement system and the unconstrained emission rate. Where a plant operates in this very wide range depends on several factors: the characteristics of the abatement system, the frequency of surveillance, and the time required to correct a violation. The last two factors are largely under the control of the agency. Average emissions and expected time in violation can be reduced both by increasing the surveillance frequency and by reducing the lag between discovery and correction of a violation.

---

[10] This is one difference between violations of pollution regulations and forms of corporate misconduct such as bribery or price fixing. In the latter, the only personnel who must be informed are top executives, who presumably share a common interest in secrecy. Deliberate violation of a pollution regulation, on the other hand, often will be impossible to keep from a large number of workers.

A Simple Model

A rather simple model of voluntary compliance can be devised and will exhibit some less-than-perfect degree of compliance on the part of the average source. No penalties are levied, but an effort to comply is *assumed* on the basis of the exogenous incentives just discussed. In this respect the model seems to mimic observed experience better than the explicit penalty models. In the following pages a series of such models is explored beginning with an extremely simple version and moving to more complicated ones. The implications of the model's structure for the frequency of violations of emission standards, and for the relationship between observable quantities such as the frequency of violations during inspections and unobservables such as the fraction of the total time in violation, will be of special interest.

The algebra underlying the model results presented in this chapter is set out in more detail in appendix 4-A. For purposes of interpreting the results, however, the following definitions and concepts are sufficient. First, for the typical source, there exist four possible states of the world, one and only one of which may hold at any one instant in time.

I_0V_0: The source is in compliance and is not under surveillance.
I_0V_1: The source is in violation and is not under surveillance.
I_1V_0: The source is in compliance and is under surveillance.
I_1V_1: The source is in violation and is under surveillance.[11]

Corresponding to these states of the world there are four probabilities, defined as the long-run fractions of the time the source is in each state. Thus:

P_{00} = The long-run fraction of the time the source is in compliance and not under surveillance.

.
.
.

P_{11} = The long-run fraction of the time the source is in violation and under surveillance.

Transition between the four states is assumed to be controlled by exponential processes. Under this assumption, the probability of moving

---

[11] This classification is similar to that of Linder and McBride (1984).

from one state, i, to another, j, in the next very short time interval is $z_{ij}\Delta t$, where z_{ij} is the key process parameter. There will be a number of z_{ij} defined in what follows, but all share some fundamental characteristics. Most important,

$$0 \leqq z_{ij} < \infty \quad \text{and} \quad 0 \leqq z_{ij}\,\Delta t \leqq 1$$

And as z_{ij} goes to zero, the probability of a transition from i to j in the next instant goes to zero; while as z_{ij} grows without limit, the probability of a transition approaches one. The common sense of this sort of process is that the average time until a transition varies inversely with z_{ij}, so that as z_{ij} goes to zero the mean time to transition grows without limit, and as z_{ij} grows without bound, transition becomes, on average, more and more nearly instantaneous.

An initial version of the voluntary compliance model is shown schematically in figure 4-2. The arrows indicate the directions of the possible transitions and the small letters are the exponential process parameters governing the transitions. Thus, transition from compliance to violation is governed by a process with exponential parameter p. Transition from no inspection to inspection is governed by the parameter q. Transition from inspection back to no inspection when no violation is under way is governed by the parameter r, and the parameter s defines the process of transition from violation with inspection back to compliance without inspection.

The assumption that the transition from compliance to violation can take place during an inspection (I_1V_0 to I_1V_1) may seem odd, but it is necessary if the position that manipulation of the timing of "failures"

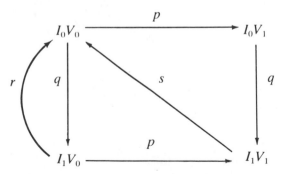

Figure 4-2. State and transition parameters in a simple voluntary compliance model

is not within the source's power is to be maintained. As it happens, however, very little of substance changes when this strict view is relaxed to allow that violations will happen only when inspections are *not* in progress. This slightly simpler case is introduced later and used as the basis for a model with differentiated inspection parameters q, depending on whether a violation is or is not in progress.

The exponential process parameters may be connected to the long-term average times spent in the several states using the method described in appendix 4-A. Again summarizing for expositional convenience, it is merely asserted here that the connection may be established via a set of linear equations. For the above model, these may be written as follows:

$$
\begin{bmatrix}
p & -q & 0 & 0 \\
q & 0 & -(r+p) & 0 \\
0 & q & p & -s \\
1 & 1 & 1 & 1
\end{bmatrix}
\begin{bmatrix}
P_{00} \\
P_{01} \\
P_{10} \\
P_{11}
\end{bmatrix}
=
\begin{bmatrix}
0 \\
0 \\
0 \\
1
\end{bmatrix}
\tag{4}
$$

where the first three equations define the relationships among the transition parameters and state probabilities, and the last equation requires that those probabilities sum to one.

The solution to the equations is given by:

$$
P_{00} = \frac{q(r + p)\,s}{B} \tag{5}
$$

$$
P_{01} = \frac{p(r + p)s}{B} \tag{6}
$$

$$
P_{10} = \frac{q^2 s}{B} \tag{7}
$$

$$
P_{11} = \frac{pqr + p^2 q + pq^2}{B} \tag{8}
$$

where $B = qs(r + q) + pr(s + q) + p^2(s + q) + pq(s + q)$

In order to begin the exploration of this model at the simplest level, consider first what happens if both r and s are assumed to grow without limit—that is, as the returns from $I_1 V_0$ to $I_0 V_0$ and from $I_1 V_1$ to $I_0 V_0$

are assumed to become instantaneous. The results for the state probabilities are:

$$P_{00} = q/(p + q) \tag{9}$$

$$P_{01} = p/(p + q) \tag{10}$$

$$P_{10} = 0 \tag{11}$$

$$P_{11} = 0 \tag{12}$$

Thus, in the long run, the fraction of time in compliance approaches $q/(p + q)$, while the fraction of time in violation approaches $p/(p + q)$. The fraction of time during which inspections are going on approaches zero. This latter result may seem odd, but is a result of assuming instantaneous returns from the inspection states to the no-inspection states. At any one point in time there is still a nonzero probability of an inspection occurring in the next instant, a probability governed by the parameter q. Furthermore, it is possible to calculate a probability that when an inspection occurs a violation will be occurring; that is, that the transition to violation will occur before the transition to inspection when the process is viewed as starting in $I_0 V_0$. That probability, as shown in appendix 4-A, is $p/(p + q)$, or the same as the long-run percentage of the time the average source spends in violation. This result is of some interest because it is possible to count the fraction of inspections that produce violations. In the simple model this fraction is an unbiased predictor of an unobservable but important number, the fraction of time the average source spends in violation. This neat connection, however, does not survive the introduction of complications, as will be seen below.

One final observation about the simplest model is that as q goes to infinity (the probability of an inspection approaching one), the fraction of time spent in violation goes to zero, while the fraction of time spent in compliance goes to one; this is perhaps not a surprising result, but again one that does not survive the introduction of complication.

The first important complication to be explored is the relaxation of the assumption of instantaneous returns to $I_0 V_0$ from the inspection states, $I_1 V_0$ and $I_1 V_1$ (the assumption that r and s are very large). Such returns do in fact take time. Some lag is implied simply by the process of issuing a notice of violation. Delay also will be incurred if the plant must order parts or make repairs. There will be some delay in returning to $I_0 V_0$ even if no violation is found. This will reflect the process of dismantling testing equipment and having a conference on the results.

Therefore, r and s must be finite, with r presumably greater than s. Then equations (5)–(8) define the long-run probabilities of being in the states.

Notice that in this more complicated model, the fraction of time spent in violation does not go to zero as the inspection parameter, q, grows without limit. Rather, the following situation obtains:

$$\lim_{q \to \infty} P_{00} = 0 \text{ (in compliance; not under surveillance)} \tag{13}$$

$$\lim_{q \to \infty} P_{01} = 0 \text{ (in violation; not under surveillance)} \tag{14}$$

$$\lim_{q \to \infty} P_{10} = \frac{s}{s + p} \text{ (in compliance; under surveillance)} \tag{15}$$

$$\lim_{q \to \infty} P_{11} = \frac{p}{s + p} \text{ (in violation; under surveillance)} \tag{16}$$

So, constant surveillance would lead to the source being in violation a fraction of the time, with the size of the fraction determined by the relative size of the violation (failure) parameter, p, and the speed of returning to compliance after violation discovery, captured by s.

Estimating time not in compliance.   A more interesting question concerns the relationship between the observables—especially the fraction of inspection *visits* producing violations but also the fraction of *inspection time* during which a source will be in violation—and the unobservable performance indicator, the fraction of *total time* in violation. To explore this question, some new notation will be useful:

VF, the fraction of *visits* producing violations
$T_I F$, the fraction of *inspection time* during which violations occur
$T_T F$, the fraction of *total time* during which violations occur.

These are derived in reverse order.

$T_T F$ in the long run is simply the sum of the probabilities, P_{01} and P_{11}, of being in cells exhibiting violation.

Thus, $T_T F = P_{01} + P_{11} = [p(r + p)s + pqr + p^2 q + pq^2]/B$ $\qquad$ (17)
$$= [pr(s + q) + p^2(s + q) + pq^2]/B$$

$T_I F$ is the fraction of inspection *time* during which violations occur; that is, in the long run the fraction P_{11} of the total inspection fraction, $P_{10} + P_{11}$.

Thus, $T_I F = P_{11} / (P_{10} + P_{11})$
$$= \frac{pqr + p^2q + pq^2}{q^2s + pqr + p^2q + pq^2} \tag{18}$$
$$= \frac{pr + p^2 + pq}{qs + pr + p^2 + pq}$$

VF is a little harder to derive for it involves "counting" average "visits" per period by calculating total length of time per period in a particular state and the average length of each visit. In particular, it is necessary to count the visits per period made to the states including inspection, I_1V_0 and I_1V_1. The fractions of a period spent in each of these states are P_{10} and P_{11} respectively. The time spent in an *average* stay in a state is estimated by the reciprocal of the exponential parameters connected with leaving it. Thus, the expected duration of a stay in I_1V_0 is $1/r + p$; and that for I_1V_1 is $1/s$. The expected *number* of visits per period is given by the total fraction of the period spent in the state divided by the expected duration of each visit. Thus:

Number of visits to state $I_1V_0 = P_{10} \left(\dfrac{1}{r + p} \right)^{-1} = (r + p)P_{10}$

Number of visits to state $I_1V_1 = sP_{11}$

Total number of inspection visits $= (r + p)P_{10} + sP_{11}$

Therefore, the fraction of monitoring visits on which a violation is detected is given by:

$$VF = \frac{sP_{11}}{(r + p)P_{10} + sP_{11}}$$
$$= \frac{s(pqr + p^2q + pq^2)}{(r + p)q^2s + s(pqr + p^2q + pq^2)} \tag{19}$$
$$= \frac{pr + p^2 + pq}{qr + qp + pr + p^2 + pq}$$

In order to compare these fractions it is necessary to make some assumption about the relative sizes of the exponential parameters. The most obvious such assumption was already alluded to: that $r > s$; that is, the speed of return to no-inspection is faster when no violation has been discovered than when one has been discovered (when the violation must be corrected simultaneously with removal of inspection). On this basis it is possible to show that $T_I F > T_T F$, the fraction of inspection time in violation is larger than the fraction of the total time in violation.[12] It is not possible, however, to derive an unambiguous relation between $T_T F$ and VF, though this would be the more interesting to have, since the fraction of visits producing violations is, realistically, the more easily estimated quantity.[13] But unfortunately, whether $T_T F$ is greater than or less than VF will depend on the magnitudes of the parameters: p, r, and s.[14]

How visit frequency affects time not in compliance. It has already been seen that $T_T F$ approaches $p/(s + p)$ as q goes to infinity. When $q = 0$, $T_T F = 1$; so that without inspection there is no compliance. Not surprisingly, it turns out that $\partial (T_T F)/\partial q < 0$ so long as p, q, r, $s > 0$. Since q, the inspection frequency parameter, is the key policy variable open to manipulation by the responsible agency, it will be valuable to

---

[12] This result is obtained via the trick of first multiplying through the numerator and denominator of $T_I F$ by $s + q$. Then,

$$T_I F = \frac{pr(s + q) + p^2(s + q) + pq(s + q)}{qs(s + q) + pr(s + q) + p^2(s + q) + pq(s + q)}$$

But because $r > s$ (so $r+q > s+q$) and $p,q,s > 0$, we have

$$T_I F > \frac{pr(s + q) + p^2(s + q) + pq^2}{qs(r + q) + pr(s + q) + p^2(s + q) + pq(s + q)} = T_T F$$

[13] A similar trick to that described in the previous footnote carries one tantalizingly close to a relationship insensitive to the relative sizes of the other parameters. But, in the end, there is an extra term in the numerator that can be removed only by subtraction, an operation having the wrong effect on the inequality.

[14] In fact, it can be shown that the relationship is as follows: If:

$$s = r + p, \; T_I F = V_F$$
$$s > r + p, \; T_I F < V_F$$
$$s < r + p, \; T_I F > V_F$$

In the limit, as $q \to \infty$, the relationships become neater. Thus,

$$\operatorname*{Lim}_{q \to \infty} (T_I F) = \frac{p}{p + s}$$

$$\operatorname*{Lim}_{q \to \infty} (T_T F) = \frac{p}{p + s}$$

$$\operatorname*{Lim}_{q \to \infty} (VF) = \frac{p}{r + 2p}$$

so that $T_I F = T_T F > VF$, when $r > s$.

explore the rate of change of time in violation with changes in q. Furthermore, to help understand what is going on in more practical terms, the variation in the *number* of monitoring visits (and the fraction of them turning up violations) with variation in q for different values of the other parameters should be explored. This is done by numerical example.

The specific values of the parameters used in these examples were derived as follows. First, in chapter 2, it was found that surveillance visits had a duration of from several hours to several days. The assumption $r = 100$ corresponds to an expected duration of about two working days. (Recall that the duration of an average stay in a state is the reciprocal of the exponential parameters connected with leaving it. One over $r(=0.01)$ as a fraction of a working year is about two days.) Because this parameter is so much greater than the others, results are insensitive to it. Second, if a violation is discovered, the source is generally ordered to return to compliance within two weeks to a month, implying values of s roughly from 10 to 25.

Values of p, which measure the reliability of the abatement equipment, vary enormously. For example, a recent survey of baghouse filter operators (Reynolds and coauthors, 1983) revealed that 42 percent of utility respondents reported 0 to 0.5 percent bags failed in the prior year, while 27 percent reported 5 percent or more bags failed[15] (see appendix 4-B). These percentages translate into values of p of about 3 and about 50, respectively. (For industrial respondents the corresponding figures were 24 percent and 38 percent.)

Overall, the parameter assumptions were: "best case" (conducive to compliance) values of $r = 100$, $s = 25$, and $p = 3$; "worst case" (conducive to violation) values of $r = 100$, $s = 10$, and $p = 50$; and an intermediate case in which $r = 100$, $s = 15$, and $p = 10$.

In table 4-2 the results of some illustrative calculations of $T_T F$, VF, and N, the number of inspection visits per period, are shown for the three cases and five values of q (the inspection probability parameter) within each case. The first case is the most favorable to compliance, with a relatively low p and relatively high s (low probability of failure and rapid return to compliance after violation detection). The second case is the least favorable, with the opposite arrangement of relative values of p (high) and s (low). And the third case represents a middle ground between one and two, with both p and s at intermediate values.

Perhaps the most striking feature of table 4-2 is the differential sen-

---

[15] A "baghouse" is an air pollution control device in which stack gases are diverted through a set of fabric filters roughly resembling giant vacuum cleaner bags and producing roughly the same effect.

Table 4-2.   Effects on Actual Compliance, Observed Violations, and
Number of Monitoring Visits Per Period of Changing the Inspection
Frequency Parameter

	$q = 1$	$q = 2$	$q = 5$	$q = 10$	$q = 20$
Case I $p = 3, r = 100,$ $s = 25$					
Fraction of time in violation, $T_T F$	0.75	0.61	0.41	0.30	0.22
Fraction of trips turning up violations, VF	0.75	0.60	0.39	0.25	0.15
Number of trips per period, N	0.98	1.9	4.6	8.7	15.8
Case II $p = 50, r = 100,$ $s = 10$					
Fraction of time in violation, $T_T F$	0.98	0.97	0.96	0.94	0.91
Fraction of trips turning up violations, VF	0.98	0.96	0.91	0.84	0.74
Number of trips per period, N	0.92	1.7	3.5	5.6	8.2
Case III $p = 10, r = 100,$ $s = 15$					
Fraction of time in violation, $T_T F$	0.92	0.86	0.74	0.64	0.56
Fraction of trips turning up violations, VF	0.91	0.84	0.68	0.52	0.40
Number of trips per period, N	0.95	1.8	4.1	7.4	12.9

sitivity of violations and visits to the inspection parameter, q, depending
on the sizes of p (violation parameter) and s (return to compliance and
no-inspection parameter). When p is small (breakdowns infrequent),
increasing q from 2 to 10, for example, decreases $T_T F$ by 50 percent.
When p is 10, the corresponding change in $T_T F$ is 25 percent. And when
p is 50, the worst case in the table, the change in $T_T F$ corresponding to
an increase in q from 2 to 10 is about 3 percent.

A second feature of interest is that sensitivity to p appears to be very
great in the low end of its range. As p goes from 3 to 10, about 15
percent of the total range from 3 to 50, the value of $T_T F$ for $q = 5$
climbs from 0.41 to 0.74, or 60 percent of its range in that case. This
effect is even greater when $q = 1$, and is much reduced when $q = 20$.
One lesson of these calculations, then, would seem to be that for voluntary compliance to work, in the sense of producing acceptable levels
of time in violation, the reliability of control equipment is crucial. If
control equipment is subject to very frequent breakdowns, the respon-

sible agency could monitor virtually constantly and still see violation time fractions around 90 percent.

What about reality? According to data presented in the previous chapter, inspections typically occur once or twice a year. For small values of p ($p = 3$) this corresponds to a violation rate of 61 percent and an observed rate, VF, of 60 percent. In appendix 4-C, some data from New Mexico are presented indicating that at typical inspection rates *per source,* the fraction of inspections turning up violations is in the neighborhood of 30 to 40 percent. Thus, although these calculations are only meant to be illustrative, and the data from enforcement inspections must be treated with skepticism, there is some reason to think that the model as structured above is unduly pessimistic.

A More Complex Model

One possible explanation for the above discrepancy is that the occurrence of a violation may change the probability of inspection, because the agency will often have, and act on, some prior information, such as a complaint or visible emissions.[16] In this case the transition probabilities are as in figure 4-3. (The path from $I_1 V_0$ to $I_1 V_1$ has been dropped to simplify the algebra. Nothing of even modest importance is lost by this simplification.)

The steady-state probabilities satisfy the following system of equations:

$$\begin{bmatrix} p & -q_1 & 0 & 0 \\ 0 & q_1 & 0 & -s \\ q_0 & 0 & -r & 0 \\ 1 & 1 & 1 & 1 \end{bmatrix} \begin{bmatrix} P_{00} \\ P_{01} \\ P_{10} \\ P_{11} \end{bmatrix} = \begin{bmatrix} 0 \\ 0 \\ 0 \\ 1 \end{bmatrix} \qquad (20)$$

The solution of which is

$$P_{00} = \frac{q_1 rs}{A}, \quad P_{01} = \frac{prs}{A}, \quad P_{10} = \frac{q_0 q_1 s}{A}, \quad P_{11} = \frac{pq_1 r}{A} \qquad (21)$$

where now

$$A = q_1 rs + prs + q_0 q_1 s + pq_1 r$$

---

[16] This model might also be applied to continuous monitoring (CEM), where now q_0 and q_1 are interpreted as periodic inspections of CEM equipment and the time required to order the plant back into compliance after receiving a periodic report indicating a violation.

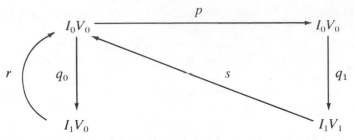

Figure 4-3.   A model with differential inspection parameters

Unfortunately this model does not do a better job of fitting the empirical data. The ratio of violations discovered to total inspections is again

$$VF = \frac{sP_{11}}{rP_{10} + sP_{11}} = \frac{spq_1r}{rq_0q_1s + spq_1r} = \frac{p}{q_0 + p} \qquad (22)$$

This means that it reflects only those routine inspections for which there is no prior information. Because these are limited to once or twice a year (q_0 is small), this model would predict that the overwhelming majority of inspections would disclose violations.

Nonetheless, the model is an interesting one because it does illustrate the value of being able to target inspections to known or suspected violators. The time between the occurrence of a violation and an inspection (that is, the expected residence time in state I_0V_1 per visit) is $1/q_1$. As $1/q_1 \rightarrow 0$, however, the number of inspections does not increase without bound, and often remains small. In fact,

$$\lim_{q_1 \to \infty} N = \lim_{q_1 \to \infty} [rP_{10} + sP_{11}] = \frac{rq_0s + prs}{rs + q_0s + pr} \qquad (23)$$

Using the parameter values of Case I from table 4-2, and letting $q_0 = 1$, the value of N is found to be 3.5 visits per year, and the values for the state probabilities are:

$P_{00} = 0.885$ (in compliance; not under surveillance)
$P_{01} = 0$ (in violation; not under surveillance)
$P_{10} = 0.0088$ (in compliance; under surveillance)
$P_{11} = 0.106$ (in violation; under surveillance)

Finally, $T_TF = P_{01} + P_{11} = .106$

Compare this to Case I, table 4-2, where inspections are at random times. When $q=5$, the expected number of inspections is roughly the same, 4.6, and yet $T_T F = 0.41$. This effect is accentuated at smaller values of p and attenuated at larger values; that is, very reliable equipment makes prior information on violation status more valuable.

Speeding Up the Return to Compliance

Detection of a violation is only half the problem, of course. A violator must also return to compliance, and the speed at which this is accomplished depends on the parameter s. (The average time from discovery of a violation to a return to compliance is $1/s$.) Up to now s has been considered as fixed or at any rate beyond the control of the agency. To examine the joint effect of q and s on the rate of violation, the path from $I_1 V_0$ to $I_1 V_1$ is again ignored. In this case, if r is assumed to become infinitely large, it can be shown that:

$$T_T F \equiv T = \frac{p(q + s)}{qs + ps + pq} \tag{24}$$

With only q to vary, and p fixed, the agency cannot make T arbitrarily small. But if s can be controlled as well, q and s together *can* be chosen to make T as small as desired. Unfortunately, this fact may not be of much practical significance, because for given T and p the required q and s may be impossibly high. To see how high, solve (24) for q in terms of T and s. Then if, for example, it is desired to maintain T equal to 0.1, while p is equal to 50, the following must be true:

$$q = \frac{ps(1 - T)}{Ts - p(1 - T)} = \frac{50\, s(0.9)}{0.1s - 50(0.9)} \tag{25}$$

For s less than 450, q will be negative; a nonsense result. For $s > 450$ but close to that value q must be very large. Thus, for $s = 460$, $q = 20,700$ must be true. While for larger s, say 4,500, q can be smaller; in this case 500. In either case, maintaining T, the fraction of the time the average source spends in violation, at one tenth requires extremely rapid returns to compliance and very high probabilities of inspection, neither of which is reasonable to expect.[17] The situation is less extreme when

---

[17] If the average source is out of compliance 10 percent of the time, 10 percent of the sources on average will be out of compliance at any one time. This way of expressing the result will be returned to in chapter 7.

$p = 3$ is assumed. Then s need only be greater than 27 to produce a positive q and for $s = 50$, $q = 59$ will be a sufficiently high inspection frequency parameter. Nonetheless, this still amounts to inspection every three or four days on average—much higher than observed rates or rates likely to be supportable by agency budgets. This theme is returned to briefly below and at much greater length, in the context of a different model, in chapter 7.

Self-Returns Compliance

Up to now it has been assumed that a source in violation would return to compliance only after the agency discovered the violation. It is possible to relax this assumption as well, to allow for the possibility that a source will return to compliance on its own. In a penalty-driven enforcement system such behavior would be commonplace, because there would be some value in avoiding detection. Under voluntary compliance, avoidance of detection has much less value, if any, since one is penalized only for failure to return to compliance. If this is so, why would a source ever decide to return to compliance on its own?

One answer is that violations may be more convenient to correct at certain times, especially during scheduled shutdowns. For example, electrostatic precipitators are divided into a number of independently energized sections. The large majority of electrostatic precipitator malfunctions are the result of broken electrodes, and repair requires shutting off the power to that section of the precipitator, reducing collection efficiency accordingly. To restore the section to service, the precipitator must be shut down and allowed to cool for several hours before the repair work can begin (Lynn and coauthors, 1976). One might expect a source to prefer to do this on its own schedule and not be forced to interrupt a production run simply to repair the abatement equipment.

On the other hand, most excess emissions from baghouses are the result of broken bags, replacement of which requires a section of the baghouse to be taken off line. This usually can be done without shutting down the entire facility, and a recent survey of baghouse operators disclosed that among utilities 78 percent of those responding replaced faulty bags immediately, while another 17 percent replaced them during scheduled outages. Among industrial users, 59 percent replaced broken bags immediately and another 14 percent waited until scheduled outages (Reynolds and coauthors, 1983). Thus, this survey suggests that baghouse users usually correct violations before discovery by the agency.

For this case, the state transition diagram is as shown in figure 4-4, where t is the self-return rate parameter.

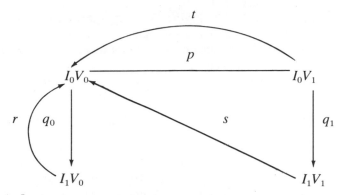

Figure 4-4. A model including self-returns to compliance

The state probabilities are now

$$P_{00} = \frac{(q_1 + t)rs}{C}, \quad P_{01} = \frac{prs}{C}, \quad P_{10} = \frac{q_0(q_1 + t)s}{C}$$

$$P_{11} = \frac{pq_1r}{C} \tag{26}$$

where $C = (q_1 + t)rs + prs + q_0(q_1 + t)s + pq_1r.$ (27)

It is probable that $t \leqq s$, that is, the time required for the source to return to compliance on its own is at least as great as the time required when it is being prodded by the agency. The models discussed earlier are special cases of this model, with $t = 0$.

Allowing $t > 0$ has two effects on the state probabilities, compared with the earlier models. First, the time spent in violation is reduced, relative to the earlier models with the same parameter values. This is evident from inspection of the expressions for P_{01} and P_{11} in (21) and (26). Thus, P_{01} and P_{11} have the same numerators as in the previous model, but the common denominator, C, is necessarily larger, since it contains additional positive terms, t being necessarily greater than zero. In addition, changing surveillance frequency q has a smaller effect on compliance than before.

Conclusion

This chapter began with a review of the economic and policy literatures dealing explicitly with problems of monitoring and enforcement in the pollution control setting. The review was organized around three key alternatives available to modelers for characterizing the situation. One of these had to do with the agency—whether it would in fact monitor or not. The other two concerned the sources—whether or not they would cheat if it were in their interest to do so; and whether or not they could control their discharges exactly or only "draw" them from a random distribution. For the most part the literature of interest assumed that the agency would monitor, but there were interesting differences in the sophistication of models on the basis of alternative choices over the second two alternatives. The review characterized models that assumed non-random discharges and no source cheating as least sophisticated and models that assumed random discharges with a possibility of willful cheating as most sophisticated. The latter models also represent the agency's problem in its toughest and most realistic light.

Actual practice of state agencies, however, as reported in chapter 2, seems to place heavy reliance on the assumption that, in fact, sources will try to comply. Heavy penalties for noncompliance are avoided, and a common approach, referred to as voluntary compliance, essentially amounts to a zero penalty. Sources discovered out of compliance are notified of the fact and given a chance to regain compliance status without a fine. Only if recalcitrance seems to be a problem—in other words if willful violation is "proved" by the source's attitude and lack of action—is a penalty approach pursued.

This observation suggested that an exploration of models designed to illuminate and ultimately to suggest solutions for the agency's problem might well begin with a model assuming an attempt to comply and disregarding the statistical inference difficulties raised by random discharges (or monitoring instrument error). The second half of the chapter set out a series of such models of increasing complexity. The voluntary compliance regime was modeled as a Markov process with four states representing the possible combinations of the alternatives in compliance/in violation and being inspected/not being inspected. Violations are produced by random equipment failure, not by willful cheating. Under these assumptions it was shown that some degree of compliance can be achieved without penalties, and the compliance status of any source can be improved with more frequent surveillance. However, the effectiveness of voluntary compliance for a particular source depends greatly on the reliability of the source's abatement equipment. Systems that break down frequently are almost always in violation, and their performance is not much improved by more frequent surveillance. As surveillance frequency increases, a practical lower limit on time in violation is implied by the delay between the discovery of a violation and the return to compliance. If the agency also can influence this time interval it can in theory achieve an arbitrarily high level of performance from any source. However, the inspection frequencies required can be extremely high, and the delays in returning to compliance impossibly short.

If, instead of making inspections randomly, it were possible to target most inspections to violations detected or suspected via some other means, much higher levels of performance would be possible for a given number of inspections. Again, the difference in performance depends on the reliability of the source's abatement equipment. For sources performing poorly, prior information on violations will not make much difference; such sources will remain out of compliance almost all the time.

The more quickly sources are assumed to return to compliance, without necessarily being forced to, the less effect increasing the frequency of inspection has on the degree of compliance. Nonetheless, the effect of assuming self-correction of violations is salutary, as one would expect.

Overall, this model suggests that voluntary compliance can work, in the sense that some degree of compliance with regulations can be attained. But it probably does not work very well for "problem" sources, which can remain out of compliance for long periods of time without apparent adverse consequences. Such sources certainly exist. For example, case studies of several in New Mexico were described in Harrington (1981).

APPENDIX 4-A

The Mathematics of the Voluntary Compliance Model

Assume that both violations and inspections are generated by exponential processes. Let the key parameters be p and q respectively. This means that the probability of a violation occurring in the next arbitrarily short time interval Δt is approximately $p\Delta t$, and the probability of an inspection being performed in such an interval is $q\Delta t$.[1] $0 \leq p, q \leq \infty$ must be true.

---

[1] The models developed here may be thought of as applications of reliability models as described, for example, in Blake, 1979, pp. 357–362.

The probability density functions for violations and inspections respectively for time t in the future are, then:

$$v(t) = pe^{-pt} \tag{A-1}$$

and

$$s(t) = qe^{-qt} \tag{A-2}$$

For a source and agency satisfying these assumptions, the mean times to first violation (T) and first inspection (S), starting from a "zero" position of compliance and no inspection are respectively:

$$E(T) = \int_0^\infty tpe^{-pt}dt = \frac{1}{p} \tag{A-3}$$

$$E(S) = \int_0^\infty tqe^{-qt}dt = \frac{1}{q} \tag{A-4}$$

Thus, as p grows without limit, the probability of a violation in the *next* very short time interval, Δt, approaches 1, and the expected time to the first violation approaches zero. This is similarly true for the inspection phenomenon.

Consider the more complicated question: what is the probability that the first violation will take place before the first inspection and thus be discovered? To provide a basis for answering it, begin by defining $u \equiv S - T$, the difference between the time to first inspection and the time to first violation. If $u > 0$ then the inspection takes place after the violation begins, and the violation is discovered.[2] If $u < 0$, when the inspection takes place there is no violation going on, and the process is begun again; that is, time is taken to be zero again and the same (timeless) probabilities of violation and inspection apply.

To find the probability that $u \geq 0$ it will be necessary to have a density function for u. Label this $h(u)$. This density function must be defined for both $u \geq 0$ and $u < 0$. In the former case observe that two events are involved: violation at some time, t, and inspection at some later time $t + u$. For $u < 0$, the sequence may be thought of as inspection at t and violation at $t - u$. But since these pairs of events can occur around any instant t, from 0 to $t \rightarrow \infty$, the probability density functions

---

[2] Note that certainty of detection is assumed, in the sense that if a violation is occurring when an inspection is made, the violation will be discovered with probability one. On the other side of the coin, if compliance is occurring at the time of an inspection, there will never be an incorrect signal of a violation. These assumptions are relaxed beginning in the next chapter.

for *u* require integration over this range. Thus:

$$h(u) \equiv \begin{cases} \int_0^\infty V(t)\, S\,(t + u)\, dt & \text{for } u \geq 0 \\ \int_0^\infty V(t - u)\, S\,(t)\, dt & \text{for } u < 0 \end{cases}$$

(A-5)

Substituting for *V* and *S* the expressions for *h(u)* are obtained:

$$h(u) \equiv \begin{cases} \int_0^\infty pe^{-pt}qe^{-q(t+u)}\, dt = \dfrac{pq}{p+q}\, e^{-qu}(u \geq 0) \\ \int_0^\infty pe^{-p(t-u)}qe^{-qt}\, dt = \dfrac{pq}{p+q}\, e^{pu}(u < 0) \end{cases}$$

(A-6)

Then the probability that $u \geq 0$ is:

$$\int_0^\infty h(u) = \int_0^\infty \frac{pq}{(p+q)}\, e^{-qu}\, du = \frac{pq}{(p+q)q} = \frac{p}{p+q}$$

(A-7)

This probability also may be interpreted as the fraction of inspections, in which violations are turned up.[3] (That $p/(p + q)$ is a fraction in the interval 0 to 1 is clear.)

Another quantity of interest is the fraction of time an average source will spend in violation. To get this quantity, it is necessary to estimate the expected length (fraction of time) of violations and the average number of violations per period (usually taken to be a year). For any particular violation, the duration of its occurrence is *u*. The expected value of *u* (for $u \geq 0$) is in turn given by:

$$\begin{aligned} E(u) &= \int_0^\infty u(h(u))\, du = \frac{pq}{p+q} \int_0^\infty ue^{-qu}du \\ &= \left(\frac{pq}{p+q}\right)\left(\frac{1}{q^2}\right)(-qu - 1)e^{-qu}\Big|_0^\infty = \left(\frac{pq}{p+q}\right)\left(\frac{1}{q^2}\right) \\ &= \frac{p}{(p+q)q} \end{aligned}$$

(A-8)

because $-que^{-qu}$ goes to zero as $u \to \infty$.

---

[3] The complement, the probability that the inspection occurs before the violation, is $1 - p/p + q = q/p + q$.

The expected number of inspections per year is the inverse of the average interval between inspections, which was found above to be $1/q$. Thus, there will be on average q inspections per year, and the expected time in violation for the average source inspected will be $p/(p + q)q$. Therefore, the time per year spent in violation by the average source will be

$$q\left(\frac{p}{(p + q)q}\right) = \frac{p}{p + q} \tag{A-9}$$

which is expressed as a fraction of the year and is labelled $T_T F \equiv T$ in the text.

The first thing to notice about this result is that the two fractions (expected fraction of inspections producing violations and expected fraction of time spent in violation) are equal. Thus, the former may be used as an unbiased estimator of the latter. This result, it turns out, depends on the particularly simple assumptions behind this first model. The introduction of a few complications breaks the link as was seen in the text.

The second thing to notice is that as the frequency of inspection increases without limit ($q \to \infty$) the time spent in violation goes to zero. Again, this proposition is only true in this very simple model and must be modified when a bit more realism is introduced.

Allowing for a Time Lag in Returning to Compliance

One of the simplifying assumptions implicit in the model discussed above is that a plant in violation returns to compliance immediately after the violation is discovered. Of course, this is unrealistic; it ordinarily takes some time just for the company to be issued a notice of violation. It is plausible to assume that this delay is also a random variable. To incorporate this possibility of delayed compliance, it is no longer possible to use the simple model discussed above.

To capture the more complicated situation it will be convenient to apply the simple theory of stochastic processes as described in Feller (1968, chapter 17), noting that the conjunction of the plant's compliance status and the agency's enforcement activity is completely described by the four events defined in the text. The transitions among these states were shown schematically in figure 4-2 and all are assumed to depend on exponential probability processes, as did inspection and violation. The parameter r applies to the process of an inspection being terminated when no violation is found. (If the source is being inspected at time t and is not violating, the probability that the inspection will be terminated before $t + \Delta t$ is $r\Delta t$.) Analogously s applies to the process of the detected violator returning to compliance and the agency calling off surveillance during the instant, Δt. The undetected violator has zero probability of returning to compliance in this model. (The path between I_0V_0 and I_0V_1 is unidirectional.) Finally, it is assumed that this process is stationary, that is, the coefficients do not change over time.

Now define

$P_{ij}(t)$ = probability that the system is in state I_iV_j ($i,j = 0,1$) at time t.

The aim is to compute these state probabilities.

Consider state I_0V_0. If at time t the system is in I_0V_0, the probability it will remain in I_0V_0 at time $t + h$ is

$P_{00}(t)(1 - ph - qh)$

where h is written for Δt. If the system is in I_1V_0 at time t, the probability it will be in I_0V_0 at $t + h$ is

$P_{10}(t)rh.$

Lastly, if the system is in I_1V_1 at t, the probability it will be in I_0V_0 at $t + h$ is

$P_{11}(t)sh.$

The probability of being in I_0V_0 at $t + h$ is then[4]

$$P_{00}(t + h) = P_{00}(t)(1 - ph - qh) + P_{10}(t)rh + P_{11}(t)sh \qquad \text{(A-10)}$$

Rearranging (A-10) we get the difference quotient

$$\frac{P_{00}(t + h) - P_{00}(t)}{h} = -(p + q)P_{00}(t) + rP_{10}(t) + sP_{11}(t) \qquad \text{(A-11)}$$

Letting $h \to 0$ in (A-11) produces the differential equation:

$$P'_{00}(t) = -(p + q)P_{00}(t) + rP_{10}(t) + sP_{11}(t) \qquad \text{(A-12)}$$

where the prime indicates the time derivative.

----

[4] This is the Chapman-Kolmogorov identity, the basic relation of the theory of stochastic processes (Feller, 1968, p. 445).

Applying this argument to the states I_0V_1, I_1V_0, I_1V_1 leads to the differential equations

$$P'_{01}(t) = -qP_{01}(t) + pP_{00}(t) \qquad \text{(A-13)}$$

$$P'_{10}(t) = -(r + p)\,P_{10}(t) + qP_{00}(t) \qquad \text{(A-14)}$$

$$P'_{11}(t) = -sP_{11}(t) + qP_{01}(t) + pP_{10}(t) \qquad \text{(A-15)}$$

The equations (A-12) through (A-15) form a system of ordinary differential equations. If the system starts in state I_mV_n the initial conditions are:

$$P_{mn}(0) = 1; \qquad P_{ij}(0) = 0 \quad \text{for} \quad (i,j) \neq (m,n)$$

These initial conditions uniquely determine the solution $\{P_{ij}(t)\}$ for a given t. This specific solution will not be sought here. Rather, the fact that all states of the system can be reached directly or indirectly from all other states can be used to determine the steady-state properties of the system independently of the starting position;[5] that is, it can be shown that the limits

$$\lim_{t \to \infty} P_{ij}(t) = P_{ij} \qquad (i,j = 0,1) \qquad \text{(A-16)}$$

exist and are independent of starting values (see Feller, 1968, p. 393). The steady state solution is found by setting the derivatives on the left hand sides of equations (A-12) through (A-15) equal to zero. The resulting system of linear equations is not of full rank, so one must be dropped. In the text A-12 was dropped, and the condition added that the probabilities of being in the four states sum to one. This produced the linear equation system set out and solved.

APPENDIX 4-B

An Example of the Relation Between Equipment Failure Rates and the Size of p: *Baghouses for Air Pollution Control*

By far the most important cause of excess emissions from a baghouse operation is bag failure. The air flowing from a blown-out bag is essentially untreated, and in a baghouse designed to achieve 99.9 percent

---

[5] The states are ergodic.

collection efficiency it does not take many such bags to have a dramatic impact on emissions.

Let B be the air flow in actual cubic feet per minute (acfm) from a blown bag, and F the total flow in acfm from the entire facility. Let d denote the design penetration (one minus collection efficiency) of the facility and d^* the penetration required by the emission regulation.

The number of bags N that fail before the facility is out of compliance must satisfy

$$\frac{BN + (F - BN)d}{F} = d^*,$$

or

$$N = \frac{F(d^* - d)}{B(1 - d)} \tag{B-1}$$

For the air flow from a blown-out bag, Lynn and coauthors (1976) give the following formula:

$$B = 2,400A(\Delta P)^{1/2} \tag{B-2}$$

where A is the cross-section area in square feet of a single bag and ΔP the pressure drop across the baghouse in inches of water.

For example, a baghouse for a 500 MW utility steam-generating station has an air flow of approximately 2.75×10^6 acfm, and requires a baghouse of perhaps 10,000, 11.5 inch-diameter bags. Assume a pressure drop of three inches of water (typical according to the survey of Reynolds and coauthors, 1983). Then, from (B-2), $B = 3,000$. Suppose $d = 0.001$ and $d^* = 0.01$. Then

$$N = 8.26 \text{ bags.}$$

The Reynolds and coauthors survey found that a good baghouse suffered bag failures in less than 0.5 percent of its bags per year. If we take 0.25 percent/year as a failure rate for a good facility, a 10,000-bag facility has 25 bag failures per year. Thus the frequency of violations is

$$p = \frac{25 \text{ bag failures/yr}}{8.26 \text{ bag failures/violation}} = 3 \text{ violations/yr}$$

At the other extreme, a poorly designed or operating facility has 4 to 5 percent bag failures per year, corresponding to a p of about 50 to 60.

APPENDIX 4-C

Some Data on Frequency of Noncompliance:
Air Pollution Sources in New Mexico

Before turning to some data from source surveillance in New Mexico, it makes sense to examine briefly the problems of making inferences regarding continuous performance on the basis of the data. Ideal information for this task is the same as the information that would be desirable from the point of view of enforcement itself: a profile of emissions over time for each source. However, such information rarely, if ever, exists for *any* source.

The kind of information that does exist depends on the source involved. For self-reporting sources the record gives thirty- or ninety-day averages of sulfur emissions, based on a calculated materials balance. For other sources there is a record of surveillance events consisting of inspections, source tests, and opacity readings.

From this record it is desirable to be able to make inferences about the continuous performance of the source, and it is worth considering for a moment what inferences are possible. First note that except for woodwaste burners, where compliance is defined in opacity terms, very little information exists regarding the *degree* of compliance or violation with the regulation because source tests are so infrequent. For asphalt processors and nonmetallic mineral processors most of the information comes from inspections, the result of which is either "this source appears to be in compliance" or "this source appears to be in violation." With this information, the best that can be expected is an estimate of the frequency with which a source will be in compliance.

But is even this a reasonable expectation? Recall from the text that whether or not $V_F = T_T F$ depended on the specific parameter values p, q, r, s. More generally, when one asserts that the frequency of violation in the surveillance record is an estimator of the overall-frequency of violation by the source, one assumes tacitly that (1) conditions do not change over time; and (2) the surveillance record is a random sample of the performance of the source. Neither of these assumptions can be made with confidence. In particular, it was observed in chapter 2 that the agency often suspects a violation before an inspection and that a source that knows about an inspection or source test in advance may act differently than it would otherwise. Both are reasons to expect bias, albeit with opposing sign.

Nonetheless, suppose that these two assumptions (no change over time and the randomness of the surveillance record) are met. It is still the case that the estimate of the frequency of violation is for most sources surprisingly imprecise, owing to the relatively small sample sizes.

Each surveillance event provides a reading of whether a source is or is not in compliance. Assuming that source behavior is stochastic, as in the model in the chapter, the surveillance event is a bernoulli random variable (where 0 represents "compliance" and 1 represents "violation"). Suppose for a particular source there exists a record of n inspections with outcomes $X_1, X_2, \ldots, X_n$. An unbiased estimator for the frequency of violation V_F is, as one might expect:

$$\hat{V}_F = \frac{1}{n} \sum X_i = \frac{S}{n}, \tag{C-1}$$

where S is the number of violations. The standard error of this estimate is

$$\text{s.e. } (\hat{V}_F) = \left[\frac{\hat{V}_F(1 - \hat{V}_F)}{n - 1} \right]^{1/2} \tag{C-2}$$

For example, if in five inspections two violations are observed (by no means an unusual case), then $\hat{V}_F = 0.4$, but a 95 percent confidence interval for V_F extends roughly from 0 to 0.9. In other words, whether the true frequency of violation was close to 0 percent of the time or 90 percent of the time, it would not be terribly unusual to find two violations in five inspections. This is a rather wide interval.

Source tests, opacity readings, and materials balances estimate the *extent* of compliance or noncompliance, rather than just whether a violation has occurred. This additional information usually makes possible more precise estimates of the frequency, as well as the seriousness, of noncompliance. This appendix, however, is confined to records of frequency of discovery of violations gathered by New Mexico's Environmental Improvement Agency (EIA). Despite the problems that attend making inferences from compliance records, they do provide useful information on the behavior of plants and insight into the enforcement process.

Steam-generating plants. Most of New Mexico's utility generators are gas fired. But coal-fired plants account for the overwhelming bulk of emissions of all kinds. And far and away the most important coal-fired source is the Four Corners Plant. This plant is not subject to any SO_2 regulation at present and except for unit 1 operates under a variance with respect to NO_x emissions. The particulate regulations call for emissions of no more than 0.135 lbs. per 10^6 BTU input for units 1, 2, and 3 and 0.5 lbs. per 10^6 BTU input for units 4 and 5. The results of source tests conducted by the agency in 1978 are summarized in table 4C-1. For units 1, 2, and 3 these tests showed that the units were comfortably

Table 4C-1.   Results of Source Tests for Particulate Emissions on Four
Corners Units 4 and 5 by Year

(emission standard: 0.5 lbs. particulates per 10^6 BTU input)

Year	Number of tests	Number of violations	Frequency of violation	Standard errors of frequency estimates
		Unit 4		
1974	10	4	0.40	0.16
1975	7[a]	0	0	—
1976	18	4	0.22	0.10
1977	10	4	0.40	0.16
1978	4	1	0.25	0.25
Overall	49	13	0.26	0.06
		Unit 5		
1974	10	2	0.20	0.13
1975	10	5	0.50	0.17
1976	8	4	0.50	0.19
1977	20	6	0.30	0.10
1978	2	0	0	—
Overall	50	17	0.34	0.07

[a]Results of tests for the first half of 1975 were missing from the Environmental Improvement Agency (EIA).

in compliance. On units 4 and 5, however, compliance could best be described as marginal. Early in 1974 the agency requested that the owner, Arizona Public Service, perform monthly source tests on units 4 and 5 and submit these records to the agency on a semiannual basis. Pursuant to this request the company conducted about fifty source tests on each unit between January 1974 and March 1978. For unit 4, about 26 percent of these tests showed violations. For unit 5, 34 percent of all tests showed violations.

Woodwaste burners.   Table 4C-2 shows the average performance during the five years to 1978 for five randomly chosen woodwaste burners. Taken as a group these five burners were found to be in violation about a third of the time, but within the group the variation was considerable. Individual mills were in violation from 13 to 55 percent of the time.

Asphalt processors.   Asphalt processors in New Mexico are subject to two regulations, one concerned with stack emissions of particulates and the other with fugitive dust. The fugitive dust regulation calls for the elimination of fugitive dust; however, for portable asphalt plants

Table 4C-2.   Average Performance of Five Woodwaste Burners, 1974–1978

Mill #	Number of opacity readings	Number of violations	Frequency of violation	Standard errors of frequency estimates
1	20	11	0.55	0.11
2	13	6	0.46	0.14
3	23	3	0.13	0.07
4	15	6	0.40	0.13
5	23	3	0.13	0.07
Overall	94	29	0.31	0.05

this is impossible, and the existence of a violation comes down to a judgment regarding whether the dust is excessive. Whether a plant is subject to a source test or an inspection, the result is a reading on violations of both stack emissions and fugitive dust, although after an inspection one can only assert a probable violation of the stack regulation.

The EIA's records for nine asphalt processing companies were examined, and the results of inspections and source tests for those companies are shown in table 4C-3. All the observations on one company were lumped together; no attempt was made to distinguish among the individual processing plants within the company. In the aggregate, these plants were found to be in violation of the regulations on about 40 percent of the visits made by the EIA. This is slightly higher than the frequency of violation among woodwaste burners, but it is striking that in both industries one finds so much variation among individual firms.

Table 4C-3.   Performance of Asphalt Processing Firms, 1972–1978

Firm	Number of source tests and inspections	Stack emission violations		Fugitive dust violations[a]	
		Frequency	Standard error	Frequency	Standard error
1	14	0.29	0.13	0.21	0.11
2	3	0	—	0.67	0.33
3	6	0.67	0.21	0.83	0.17
4	7	0	—	0.29	0.18
5	5	0.60	0.24	0.20	0.20
6	10	0.60	0.16	0.50	0.17
7	5	1.00	—	0.20	0.20
8	7	0.43	0.20	0.29	0.18
9	7	0	—	0.29	0.18
Overall	64	0.40	0.06	0.39	0.06

[a] The emission regulations for asphalt processing plants allow no fugitive emissions at all. However, this standard is too stringent to be taken literally, and inspectors at their discretion cite only "excessive" fugitive emissions.

Table 4C-4. Performance of Nonmetallic Mineral Processing Plants, 1972–1978

Plant	Number of source tests and inspections	Stack emission violations		Fugitive dust violations[a]	
		Frequency	Standard error	Frequency	Standard error
1	8	0.25	0.16	0.50	0.19
2	14	0.64	0.13	0.07	0.07
3	4	0.75	0.25	0.25	0.25
4	6	0.33	0.21	0	—
5	5	0	—	0	—
6	4	0.75	0.25	0	—
Overall	41	0.45	0.08	0.82	0.06

[a] The emission regulations for these plants also allow no fugitive emissions. As with asphalt processors, inspectors use their discretion and cite only emissions that appear excessive.

Nonmetallic mineral processors. Table 4C-4 gives the performance of six mineral processors in the state. This information, however, is offered only for the sake of completeness; four of these plants have been in situations in which the phrase "continuous compliance" is hardly applicable because of exceptions and variances.

References

Allingham, M. G., and A. Sandmo. 1972. "Income Tax Evasion: A Theoretical Analysis," *Journal of Public Economics* vol. 1, no. 3/4, pp. 323–338.

Beavis, Brian, and Martin Walker. 1983. "Achieving Environmental Standards with Stochastic Discharges," *Journal of Environmental Economics and Management* vol. 10, pp. 103–111.

Becker, Gary S. 1968. "Crime and Punishment: An Economic Analysis," *Journal of Political Economy* vol. 76 (March/April) pp. 169–217.

Berthouex, P. M., and W. G. Hunter. 1975. "Treatment Plant Monitoring Programs: A Preliminary Analysis," *Journal of the Water Pollution Control Federation* vol. 47, pp. 2143–2156.

————,————, and L. Pallesen. 1978. "Monitoring Sewage Treatment Plants: Some Quality Control Aspects," *Journal of Quality Technology* vol. 10, pp. 139–149.

Blake, I. F. 1979. *An Introduction to Applied Probability* (New York, John Wiley & Sons).

Bohm, Peter, and Clifford S. Russell. 1985. "Comparative Analysis of Alternative Policy Instruments," in Allen V. Kneese, ed., *Handbook of Environmental and Resource Economics* (New York, North Holland).

Braithwaite, John. 1982. "The Limits of Economism in Controlling Harmful Corporate Conduct," *Law and Society Review* vol. 16, no. 3.

Casey, Don, Peter N. Nemetz, and Dean H. Uyeno. 1983. "Sampling Frequency for Water Quality Monitoring: Measures of Effectiveness," *Water Resources Research* vol. 19, no. 5 (October) pp. 1107–1110.

Courtney, F. E., C. W. Frank, and J. M. Powell. 1981. "Integration of Modelling, Monitoring, and Laboratory Observation to Determine Reasons for Air Quality Violations," *Environmental Monitoring and Assessment* vol. 1, no. 2, pp. 107–118.

Crandall, Robert. 1983. *Controlling Industrial Pollution* (Washington, D.C., Brookings Institution).

Curran, T. C., and B. J. Steigerwald. 1982. "Data Analysis Consequences of Air Quality Measurement Uncertainty." Paper presented at the annual meeting of the Air Pollution Control Association, New Orleans, June.

Downing, Paul B., and William D. Watson, Jr. 1973. *Enforcement Economics in Air Pollution Control.* EPA 600/5-73-014 (December) (Washington, D.C., U.S. Environmental Protection Agency).

———. 1974. "The Economics of Enforcing Air Pollution Controls," *Journal of Environmental Economics and Management* vol. 1, pp. 219–236.

———. 1975. "Cost-Effective Enforcement of Environmental Standards," *Journal of the Air Pollution Control Association* vol. 25, no. 7 (July) pp. 705–710.

Feller, William. 1968. *Introduction to Probability Theory and Its Applications,* vol. 1. (New York, John Wiley & Sons).

Freeman, A. Myrick III. 1982. *Air and Water Pollution Control: A Benefit-Cost Assessment.* (New York: John Wiley & Sons).

Gordon, Glen E. 1980. "Receptor Models," *Environmental Science and Technology* vol. 14, no. 1, pp. 792–800.

Greer, Charles R., H. Kirk Downey, and Mark A. Tallent. 1978. "Compliance with EPA Rules: A Decision Criteria Model," *Industrial Relations* vol. 17, no. 3 (October) pp. 347–352.

Harford, John D. 1978. "Firm Behavior Under Imperfectly Enforceable Pollution Standards and Taxes," *Journal of Environmental Economics and Management* vol. 5, no. 1, pp. 26–43.

Harrington, Winston. 1981. *The Regulatory Approach to Air Quality Management* (Washington, D.C., Resources for the Future).

Linder, Stephen H., and Mark E. McBride. 1984. "Enforcement Costs and Regulatory Reform: The Agency and Firm Response," *Journal of Environmental Economics and Management* vol. 11, no. 4 (December) pp. 327–346.

Lynn, D. A., G. Deans, K. Hill, M. Rei, and P. Spawn. 1976. "Assessment of Particulate Attainment and Maintenance Problem" (Bedford, Massachusetts, GCA Corporation).

McInnes, Robert G., and Peter H. Anderson. 1981. *Characterization of Air Pollution Control Equipment Operation and Maintenance Problems* (Washington, D.C., U.S. Environmental Protection Agency).

McKean, Roland N. 1980. "Enforcement Costs in Environmental and Safety Regulation," *Policy Analysis* vol. 6, no. 3, pp. 269–289.

Melnick, R. Shep. 1983. *Regulation and the Courts: The Case of the Clean Air Act* (Washington, D.C., Brookings Institution).

National Academy of Sciences. 1977. *Environmental Monitoring, A Report to the U.S. Environmental Protection Agency from the Study Group on En-*

vironmental Monitoring (Washington, D.C., National Academy of Sciences, National Research Council).

Pickett, E. E., and R. G. Whiting. 1981. "The Design of Cost-Effective Air Quality Monitoring Networks," *Environmental Monitoring and Assessment* vol. 1, no. 1, pp. 59–74.

Posner, Richard. 1977. *Economic Analysis of Law,* 2d ed. (Boston, Little, Brown).

Reynolds, J., S. Kreidenweis, and L. Theodore. 1983. "Results of a Baghouse Operation and Maintenance Survey on Industry and Utility Coal-Fired Boilers," *Journal of the Air Pollution Control Association* vol. 33, no. 4 (April).

Stigler, George J. 1970. "The Optimum Enforcement of Laws," *Journal of Political Economy* vol. 78, pp. 526–536.

Storey, D. J., and P. J. McCabe. 1980. "The Criminal Waste Discharger," *Scottish Journal of Political Economy* vol. 27, no. 1 (February) pp. 30–40.

Tietenberg, T. H. 1980. "Transferable Discharge Permits and the Control of Stationary Source Air Pollution: A Survey and Synthesis," *Land Economics* vol. 56, no. 4 (November) pp. 391–416.

U.S. Council on Environmental Quality. 1979. *Tenth Annual Report* (Washington, D.C., Government Printing Office).

U.S. Environmental Protection Agency. 1978. *Civil Penalty Policy* (Washington, D.C., Office of Enforcement, April 11).

U.S. Government Accounting Office. 1979. *Improvements Needed in Controlling Major Air Pollution Sources* (Washington, D.C., U.S. GAO).

————. 1983. *Wastewater Dischargers Are Not Complying with EPA Pollution Control Permits* (Washington, D.C., U.S. GAO) GAO/RCED-84-53 (December).

Vaughan, William J., and Clifford S. Russell. 1983. "Monitoring Point Sources of Pollution: Answers and More Questions from Statistical Quality Control," *The American Statistician* vol. 37, no. 4, pt. 2 (November) pp. 476–487.

Viscusi, W. Kip, and Richard J. Zeckhauser. 1979. "Optimal Standards with Incomplete Enforcement," *Public Policy* vol. 27, no. 4 (Fall) pp. 437–456.

Watson, William D., Jr., and Paul B. Downing. 1976. "Enforcement of Environmental Standards and the Central Limit Theorem," *Journal of the American Statistical Association* vol. 71, no. 355 (September) pp. 567–574.

Wenders, John T. 1975. "Methods of Pollution Control and the Rate of Change in Pollution Abatement Technology," *Water Resources Research* vol. 11, no. 3 (June) pp. 393–396.

Williamson, M. R. 1981. "SO_2 and NO_2 Mass Emission Surveys: An Application of Remote Sensing," in Air Pollution Control Association, *Continuous Emission Monitoring: Design, Operation, and Experience* (Pittsburgh, Pennsylvania, APCA).

Witten, A. J., F. C. Kornegay, D. B. Hunsaker, Jr., E. C. Long, Jr., R. D. Sharp, P. J. Walsh, E. A. Zeighani, J. S. Gordon, and W. L. Lin. 1982. *The Implications of a Stochastic Approach to Air Quality Regulation* (Oak Ridge, Tennessee, Oak Ridge National Laboratory) ORNL/TM-8440.

Wittman, Donald. 1977. "Prior Regulation Versus Post Liability: The Choice Between Input and Output Monitoring," *Journal of Legal Studies* vol. 6 (January) pp. 193–211.

5
Statistical Background

Rather than continue to revise the structure of the voluntary compliance model, we will turn now to the problem of statistical complications. Once the twin elements of stochastic discharge patterns and discharge measurement errors have been introduced, it will be seen that the mere occurrence of an inspection (monitoring visit) while a true violation is in progress does not guarantee detection of the violation. On the other hand, a visit may turn up an apparent violation even though the source really is in compliance.

First, then, this chapter provides an informal argument showing how and why statistical problems complicate monitoring and enforcement. More technical sections follow that give key definitions and demonstrate the effect of randomness upon the analysis. The most important part of the chapter is the last section, which presents definitions of and connections between error probabilities, concepts central to a working understanding of how random variation and error influence the monitoring and enforcement problem.

On the basis of the background provided in this chapter, chapter 6 provides a model for the optimal design of a monitoring strategy when random error is taken into account. This model represents the voluntary compliance view: the sources subject to monitoring are assumed to be trying to comply (how hard is not specified); they may fail to comply but do not choose to violate; and no penalties are imposed for violations discovered. The only outcome of discovery is correction.

It will also be seen, however, that this model has its own set of inadequacies, not the least of which is that agency information requirements far exceed what is customarily or even what could feasibly be collected. Therefore, chapter 7 will explore a third model, one that allows the use of less demanding criteria for policy design, and consequently relaxation of the information requirement. Furthermore, the model in chapter 7 will allow an attack on two other features of the problem: the costs of monitoring and enforcement activity as related to an agency budget constraint; and the existence of sources that choose to violate rather than comply.

How Randomness Complicates Monitoring and Enforcement

To see how randomness complicates the monitoring and enforcement problem, consider a sequence of cases beginning with one in which randomness plays no part and moving toward the more realistic situation in which both actual discharges and agency measurement capabilities have random components.

The Nonrandom Case

In this polar case the source has perfect control over its discharge. If it chooses to emit D tons per day of a residual, that is what is emitted. If it faces an emission standard $\hat{D}$, it can choose either to comply or to violate (and the exact degree of violation). Emissions per unit time do not fluctuate around D as a mean, they equal D in every time unit.

Correspondingly, the environmental agency can determine with perfect accuracy what the firm is discharging per unit time and over the period for which the standard is set. If D_t is being emitted in time period t and the agency measures the discharge, its instruments show a reading of exactly D_t.

What is the nature of the problem in such a situation? That depends on other assumptions about the capabilities and costs of the source and agency and on the information available to the agency. One problem arises if the source has notice of every monitoring visit and can adjust from violation to compliance within the time from notice to visit. Then there is no reason for the source ever to be found in violation; but, on the other hand, there is never any reason for it to be in compliance except when the agency monitoring team is actually measuring its emissions. Here, perfect capabilities can translate into perfect cheating. Such a problem can only be remedied by invoking the mythical tamper-proof, continuous-recording monitoring instrument, thus providing a continuous agency presence.

At the other extreme, it can be assumed that, while capabilities are perfect, no notice of a monitoring visit need be given and that the source will never have time to adjust from violation to compliance before measurements are taken. A perfect remote monitoring instrument would produce such a situation. (Chapter 3 provides information and references relevant to remote monitoring technology.) Two questions immediately arise. First, since the source that chooses to violate *can* be caught, there has to be an enforcement policy for dealing with discovered instances of noncompliance. (Examples of penalties are a fixed fine per violation regardless of severity or a fixed fee per unit of violation. For some implications of the choice among these refer back to figure 4-1 in chapter 4.) The second question concerns how often to monitor. Continuous monitoring should produce, with any sufficiently high penalty, continuous compliance. But because the damages from a violation are almost certainly not infinite and because there are positive costs to monitoring as well as emission control, it usually will not be socially optimal for the agency to choose this route. Instead, some monitoring probability less than one will be chosen; and in the classical framework, source and agency will settle into a steady state, with the source picking a level of violation and the agency a monitoring probability per unit time.[1] The only uncertainty is exactly when the source will be caught—not *if* it will be caught. This is the basic model for the neoclassical economic approaches reviewed in chapter 4.

Introducing Randomness

A random element with substantial implications for the problem can enter in two ways. First, the source in fact may not have perfect control over its emissions, but instead can be thought of as drawing a discharge D_t each period from a "random discharge generator" that may be characterized by its mean, say μ_D, and its variance, say σ_D^2. Lest the reader think such a possibility is unrealistic, figure 5-1 provides an example from the quality control literature, showing daily discharge of five-day, biochemical, oxygen-demanding organics (BOD-5) from the Madison, Wisconsin, sewage treatment plant (from Berthouex, Hunter, Pallesen, and Shih, 1976, pp. 691, 692). Such variability in discharges can be caused by changes in plant inflow or the weather, the presence or absence of biocides in the waste being treated, or just plain operator errors.

---

[1] Notice that the agency must choose a probability, not a fixed schedule, when violation can be chosen by the source. Under a fixed and announced schedule, the situation returns to the other pole of the nonrandom case—no violations observed; no compliance except at monitoring times.

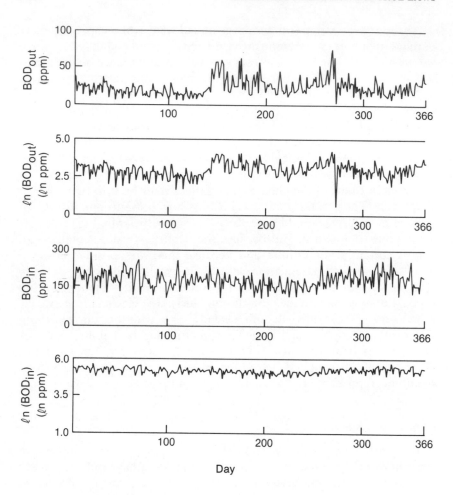

Figure 5-1.   Influent and effluent BOD-5 for the Madison, Wisconsin, sewage treatment plant during 1971.   *Source*: P. M. Berthouex, W. G. Hunter, L. Pallesen, and C. Y. Shih, "The Use of Stochastic Models in the Interpretation of Historical Data from Sewage Treatment Plants," *Water Research* vol. 10 (1976) pp. 691–692.

For other pollutants and for private firms producing products rather than only treating wastes, the list of reasons for discharge variations lengthens to include changes in the level of production, the product mix or quality, and the input quality.

In such a situation, the agency, as guardian of the public interest, faces two related problems in coming to terms with the variability. First, it must recognize such variability as a fact of life when it sets emission standards. Second, depending on how it sets those standards, it also may have to take variability into account in its monitoring and enforcement activity. The essence of the situation is this.[2]

- If the agency sets a standard and intends to enforce it as an absolute upper limit, it must recognize that the discharger who intends to comply will have to choose a treatment process such that the mean or target discharge is sufficiently far below the standard that random variation will produce violations with acceptably small probability. What is acceptable to a source will depend on the monitoring probability and the penalty structure.
- If the agency announces a discharge standard other than an absolute upper limit it must announce at the same time what the chosen characteristics are. For example, is the standard to be a mean? Or a value to be exceeded not more than 25 percent of the time? Or what? In this case, an observation of a discharge above the standard means nothing by itself. Only repeated samples allow the agency to estimate the appropriate statistic and decide whether or not there has been a violation.

But notice that in neither case can the agency be certain that the source meant to be in violation. As a general rule, after optimal adjustment to the announced standard, the source will still produce discharge patterns violating that standard with probability greater than zero, even in the absence of a will to violate. Thus, measured violations

---

[2] As a general matter, the agency may be choosing emission standards either on technological or ambient quality grounds. The former method, which is common in current U.S. environmental regulation, begins with the hypothetical application of a particular discharge reduction technology to a plant of given size in a particular industry class. The ability of the technology to reduce discharges is estimated on the basis of data from actual plants chosen because they have the equipment in place. (These classes may be very finely divided.) The result of this exercise is an achievable mean discharge rate. To this is added a factor to allow for variability of the sorts mentioned in the text, whether formally referred to as some multiple of the observed standard deviation in discharges or not. The factor is not so large as to preclude all possibility of a violation when the plant is trying to comply, and this is a source of legal argument against the whole process. (See chapter 3.)

Setting discharge standards on the basis of ambient standards in the real world of variable discharges requires in principle a prior decision about the acceptability of violation of the ambient standard. This must be translated into appropriate discharge standards for the relevant sources, taking into account their variability patterns and any correlations among them (positive or negative). This is a difficult process and is not pursued here. For some further discussion, see Beavis and Walker, 1983; and for complications, Davidson and Hopke, 1984, and Shaw and coauthors, 1984.

do not necessarily imply an attempt to cheat. The opposite is also true. The source can choose a target with the intent to violate the standard but produce monitoring results that show compliance in the short run. In whatever way the standard is defined, the agency will face residual uncertainty about the appropriateness of its enforcement decisions. Even though a violation will be well defined, intentional *violators* will be identified correctly only with some probability less than one; and similarly for sources intending to comply. Said another way, a source accused of violation will have the possibility of a defense based on its intent to comply having been thwarted by random events beyond its control.

Another cause of significant randomness is the measurement instruments used by the agency to monitor. These might give the correct answer on *average* if applied to a particular discharge level many times. But any particular reading might be in error by a random amount, itself drawn from an error distribution with zero mean and variance σ_I^2; that is, the actual reading when discharge level D_t is being measured might be $D^I = D_t + \varepsilon$, where $E(\varepsilon) = 0$ [$E(\cdot)$ is the expectation operator] and $\text{var}(\varepsilon) = \sigma_I^2$. In this situation the agency cannot be sure what the source is actually discharging. However the standard is defined, the agency cannot identify violations and compliance with certainty. Again it is worth stressing that instrument error is a real, not imagined, problem.[3] For examples of sizes of measurement errors related to determination of certain important air pollutant discharge rates, see Brenchley, Turley, and Yarmac, 1973, table 8-2, p. 163; table 8-5, p. 170; and table 15-1, p. 274.[4] In order to incorporate random discharge patterns and instrument errors into a view of the monitoring problem, it will be necessary to be somewhat more systematic in defining both the difficulties implied and the methods for dealing with them.

Some Definitions and Relationships

Bias, accuracy, and precision are three items used frequently, though by no means consistently, in discussions of stochasticity in monitoring—especially that element of stochasticity introduced by the measurement apparatus itself. It will be useful to be clear about what is meant by

---

[3] A measurement instrument also may have systematic error that causes it to give a reading consistently too high or too low. The next section in this chapter will discuss such systematic error.

[4] The handling of multiple sources of error in a single measurement sequence is discussed later in the chapter.

these terms, even if that requires picking from among several definitions to be found in the literature. The related idea of calibration also will be discussed.

Throughout this section interest will center on the relationship between the actual discharge rate D_t and the measurement D_t^I of that rate as determined by the monitoring instruments. In general there will be no instrument capable of measuring D_t directly. Rather, the instruments measure certain quantities, such as gas flow rate, mass of particulate matter on a filter, pressure drop in inches of mercury, or dissolved oxygen concentration. Let x_i^R denote the true values of such quantities, which are assumed to be related to the discharge rate by a function $f(\cdot)$:

$$D_t = f(x_{1t}^R, \ldots, x_{nt}^R)$$

Let x_1^I denote the *measured* quantities as determined by the monitoring instruments. Then the discharge measurement D_t^I is calculated as

$$D_t^I = f(x_{1t}^I, \ldots x_{nt}^I)$$

But for now, it will be most useful to concentrate on the simple situation in which a direct measurement of D is assumed to be possible.

Direct Measurement of Discharge

Let t refer to the smallest time unit of interest, say a second. Then any of the following quantities might conceivably be of interest to the responsible agency:

$$D_h \equiv \text{hourly discharge} = \sum_{t=1}^{3,600} D_t$$

$$D_d \equiv \text{daily discharge} = \sum_{h=1}^{24} D_h$$

$$\overline{D}_h \equiv \text{average hourly discharge over a day}$$

$$\equiv \sum_{h=1}^{24} D_h/24 = D_d/24$$

$$\overline{D}_d \equiv \text{average daily discharge over a (thirty-day) month}$$

$$\equiv \sum_{d=1}^{30} D_d/30$$

$D_t\text{max}_T \equiv$ the maximum per-second rate of discharge over some period T.

$D_h\text{max}_T \equiv$ the maximum hourly discharge over some period T.

.

.

.

$D_t(.05)_T \equiv$ the level of per-second discharge higher than all but 5 percent of the discharge rates over some period T.

Many other statistics based on the measurements D_t may be calculated and used by the agency for enforcement purposes.

Let the D_t be drawn, as described above in the text, from a distribution with mean μ_D and variance σ_D^2. Furthermore, for simplicity, assume that the D_t are uncorrelated over time;[5] that is,

$$E[(D_t - \mu_D)(D_{t+1} - \mu_D)] = 0.$$

Then the hourly discharge totals may be thought of as draws from a distribution with mean 3,600 μ_D and variance 3,600 σ_D^2. Analogous results hold for *total* discharges over other periods. (See, for example, Parzen, 1960, beginning on p. 366.)

The distribution of *averages* of the D_t taken over periods of seconds look quite different, however. The general formulae are as follows.[6] Let

$$S_T = \sum_1^T D_t \text{ and } M_T = (1/T)\, S_T; \text{ then } E[M_T] = \mu_D = E(D_t) \text{ and } \text{Var}(M_T)$$

$= 1/T \,\text{Var}\,(D_t) = 1/T\, \sigma_D^2$. Thus the variance of the *average* discharge over a period of T seconds falls as T increases.

Consider next the measurement of a sequence of discharges, D_t, using the assumed instrument; call them D_t^I. If the expected value of D_t^I is μ_D then the instrument is said to be *unbiased*. This is true when D_t is measured with random error having zero mean, for in that case, after T measurements it is possible to calculate:

$$M_T^I = \left(\sum_{t=1}^T D_t^I\right)/T = \left(\sum_{t=1}^T (D_t + \varepsilon_t)\right)/T$$

[5] For methods of dealing with serially correlated discharges, see Gardenier, 1982.

[6] In these terms, the results for distributions of *totals* are $E(S_T) = T\mu_D$ and $\text{Var}(S_t) = T\,\text{Var}(D_t)$.

and

$$E[M_T^I] = \frac{1}{T} \sum_{t=1}^{T} E(D_t) + \frac{1}{T} \sum_{t=1}^{T} E(\varepsilon_t) = \mu_D$$

Unfortunately, the sample variance of the measurements from one monitoring device cannot lead to an unbiased estimate of the variance of the discharges, because the measurement variance confounds the instrument variance and the discharge variance. Let

$$s_M^2 = \frac{1}{T-1} \sum_{t=1}^{T} (D_t^I - M^I)^2,$$

the unbiased estimate of the measurement variance. If the D_t are mutually and serially uncorrelated it is straightforward to show that

$$E(s_M^2) = \sigma_D^2 + \sigma_I^2, \tag{1}$$

where $\sigma_I^2 = \text{Var } \varepsilon_t$ is the instrument variance.

Suppose, on the other hand, N identical instruments are used simultaneously to measure a single D_t, the discharge in a single second ($N > 1$). Each observation is of the form $D_{tn}^I = D_t + \varepsilon_{tn}$. Then if

$$M_t^I = \sum_{i=1}^{N} \frac{D_{ti}^I}{N}$$

is the mean of the instrument measurements, it is possible to estimate the instrument variance σ_I^2 by

$$s_I^2 = \frac{1}{N-1} \sum_{i=1}^{N} (D_{ti}^I - M_t^I)^2$$

The instrument variance σ_I^2 will be called the *precision* of the instrument. If the precision is known or can be estimated, it can be substituted back into equation (1) to obtain an unbiased estimate of the variance of the discharge stream σ_D^2.

An instrument that is *biased* has a systematic measurement error built in. Then the expectation of the errors, ε, in individual measurements is not zero. The systematic errors that constitute bias can appear in different forms, but two simple examples will give the flavor of the implications for interpretation of the measurement statistics:

- Constant error, so that $D_t^I = c + D_t + \varepsilon_t$, where c is a constant
- Proportional error, so that $D_t^I = bD_t + \varepsilon_t$

In these cases the results of calculations based on reported measurements may be summarized as follows:

- Sample mean: $E(M_T^I) = 1/T \Sigma ED_t^I = c + \mu_D$
  Sample variance: $E(s_M^2) = 1/(T - 1) \Sigma (D_t^I - M_T^I)^2 = \sigma_D^2 + \sigma_I^2$
- Sample mean: $E(M_T^I) = b\mu_D$
  Sample variance: $E(s_M^2) = b^2\sigma_D^2 + \sigma_I^2$

Thus, the constant bias confounds the estimate of the mean, while the proportional bias affects estimates of both the mean and variance of the discharge stream.

Another interesting statistic can be defined if the measurement of a single value D_t is considered: the mean square error (MSE) of the estimate of D_t from an instrument with constant bias c and random error ε_t with mean zero and variance σ_I^2.

$$
\begin{aligned}
\text{MSE } (D_t^I) &= E (D_t^I - D_t)^2 \\
&= E (D_t + c + \varepsilon_t - D_t)^2 \\
&= c^2 + \sigma_I^2 \\
&= (\text{bias})^2 + \text{precision.}
\end{aligned}
$$

This measure is sometimes referred to as the *accuracy* of the estimate of D_t provided by the instrument. In this usage, accuracy is a summary measure combining precision and bias. If N identical instruments are used, the MSE of the mean estimate M_t^I is

$$
\text{MSE } (M_t^I) = c^2 + \frac{\sigma^2}{N}
$$

Thus, the use of N instruments can increase the precision of a measurement but can do nothing about the bias.[7]

---

[7] It is not quite correct to say, however, that one should be indifferent about instruments with equal accuracy but different mixes of bias and precision. If bias is relatively large, the frequency distribution of absolute errors ($D_t^I - D_t$) can be quite different from that in which bias is low relative to the variance of the instrument. Thus, on the basis of the underlying components of accuracy, one might prefer one or the other of two equally *accurate* instruments (See Cochran, 1977, p. 15).

Since bias confounds attempts to measure discharge rates even in the expectational (multiple sample) sense, it is very desirable to detect and eliminate it from monitoring instruments. This is the function of the process of *calibration*. In this process, the instrument is applied to a reference standard, and the difference between instrument reading and standard is measured.[8] If this is repeated at several levels of reference standard, the nature of the bias, if any, can be detected and a calibration curve produced. Examples of such curves are presented in figures 5-2a and 5-2b. These correspond respectively to proportional and constant bias. When such curves are available, instrument readings can be corrected to reference readings leaving only the noise represented by ε.

When Discharge Must Be Estimated from a Set of Other Measurements

As was noted at the beginning of this section, it is not usually the case that a discharge quantity corresponding to a standard can be read directly from a single instrument. More often a series of measurements must be taken and functionally combined to obtain an indirect estimate of discharge. In general, this situation may be represented by functional notation as,

$$D_t^I = f(x_{1t}^I, \ldots, x_{mt}^I)$$

where the x_{it}^I are actually measured and D_t^I is calculated using $f(\cdot)$. But this abstract formula fails to provide a feel for just how complex the relation between the measured x values and the calculated pollution loadings can be. To capture the essence of a realistic problem, consider the measurement of particulate matter in stack gas using Environmental Protection Agency (EPA) method 5. (The exposition is based on Brenchley and coauthors, 1973, pp. 215–219.)

The quantity to be monitored is particulate mass emission rate per hour, PMR. This, in turn, is the product of two other calculated quantities.

Q = stack-gas flow rate in cu. ft. per hr. (after correction for water vapor)

---

[8] The reference standard is an example, having known value, of the quantity the instrument is designed to measure. The preparation of reference standards for the many measured quantities encountered in pollution monitoring involves a variety of technical problems, a discussion of which is beyond the scope of this simple background chapter. For further detail and discussion, see National Academy of Sciences, 1977, especially vols. IV and IVa; Environmental Science and Technology, 1978; and Hunter, 1980.

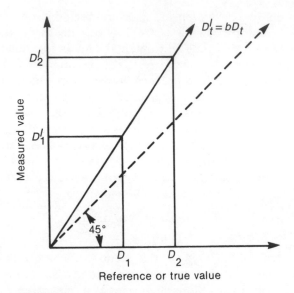

Figure 5-2a.   Proportional calibration relationship between D_t^R and D_t^I

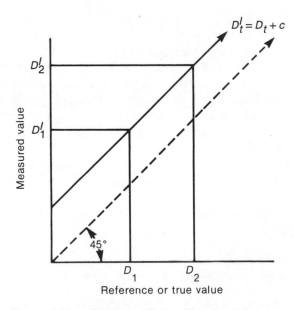

Figure 5-2b.   Constant offset calibration relationship between D_t^R and D_t^I

C = concentration of particulates in lb./cu. ft. of dry stack gas.

But Q and C cannot be measured directly either, and the actual method of obtaining an estimate of PMR involves more than a dozen separate measurements of stack-gas volumes, pressures, temperatures, and percentage composition, as well as stack-cross sectional area and the mass of collected particulates.

The sampling arrangements for making the required measurements are sketched on figure 5-3, and the quantities measured by each part of the train are indicated in parentheses. A detailed description of the required measurements and the resulting formulas are given in appendix 5-A. For purposes of the text presentation it is sufficient to say that these formulas are quite complicated.

In such contexts, techniques for the approximation of the overall results of simultaneous measurement errors are useful. The technique to be described briefly uses a Taylor-series expansion. The next section shows how the impact of systematic and random measurement errors can be approximated in complex problems like particulate loading measurement.[9]

First, consider the propagation of systematic error (bias) when $y = f(x_1, \ldots, x_m)$. Let $\{x_i^R\}$ represent the vector of true values of the independent variables, and let each x_i be measured with constant systematic error, b_i, and let the vector of biased measurements be $\{x_i^o\}$, measured without error. If the biases are known, the true values can be recovered as

$$x_i^R = x_i^o - b_i$$

and the true dependent variable y^R can be computed:

$$y^R = f(x_1^o - b_1, \ldots, x_m^o - b_m)$$

Regardless of whether the b_i are known it may be of interest to determine the contribution of the individual biases to the imprecision Δy in the estimate of y, defined by

$$\Delta y = f(x_1^o, \ldots, x_m^o) - f(x_1^R, \ldots, x_m^R)$$

The marginal effect of each b_i on Δy is then

$$\frac{\partial(\Delta y)}{\partial b_i} = f_i(x_1^R, \ldots, x_m^R)$$

[9] For a review of this and alternative methods of error propagation analysis, see Evans, Cooper, and Kinney, no date.

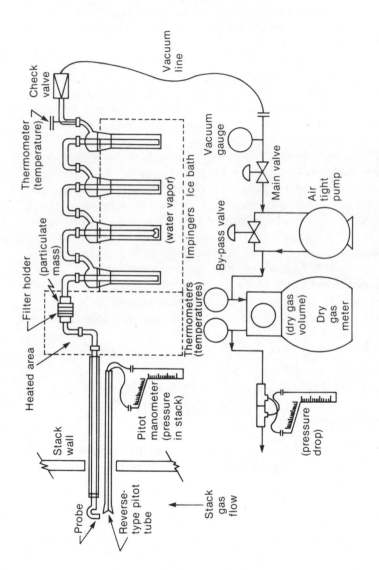

Figure 5-3. Sampling train for particulate matter (EPA method 5). *Notes:* Heated area used to prevent moisture condensation in filters measuring particulate mass; measured separately are the temperature of the stack gas and the makeup of stack gas.

The overall error then may be calculated, to a first order approximation, as

$$\Delta y \cong \sum_{i=1}^{m} b_i f_i(x_1^R, \ldots, x_m^R)$$

Next consider the propagation of random error through a functional calculation of the form $y = f(x_1, \ldots, x_m)$ with random error but no systematic error in the x_i. Let

$$x_i^o = x_i^R + \varepsilon_i, i = 1, \ldots, m,$$

where $E(\varepsilon_i) = 0$, so that $E(x_i^o) = x_i^R$. Expand f in a Taylor series around $(x_1^R, \ldots, x_m^R)$, assuming that the errors ε_i are sufficiently small so that only first order terms need be considered:

$$y^o = f(x_1^o, \ldots, x_m^o) \cong f(x_1^R, \ldots, x_m^R) + \sum_{i=1}^{m} \varepsilon_i f_i(x_1^R, \ldots, x_m^R)$$

With this assumption y^o is approximately unbiased, so that

$$E(y^o) \cong y^R$$

The variance of y^o is given by

$$\sigma_y^2 = E(y^o - y^R)^2 \cong E[\sum_{i=1}^{m} \varepsilon_i f_i(x_1^R, \ldots, x_m^R)]^2$$

Put $\sigma_{\varepsilon_i}^2 = E(\varepsilon_i^2)$ and suppose the errors among the x_i are independent (that is, $E(\varepsilon_i \varepsilon_j) = 0$ for $i \neq j$). Then

$$\sigma_y^2 \cong \sum_{i=1}^{m} \sigma_{\varepsilon_i}^2 f_i(x_1^R, \ldots, x_m^R)^2$$

Thus the variance in the measurement y depends on the variance of the x_i and the partial derivatives of the function f, evaluated at the true values x_i^R. (For further details and examples see Larsen and Marx, 1981, appendix 3-1, and Bevington, 1969.)

In parallel to the discussion in the preceding section two further results can be derived. First, measurement error is again confounded with process variation. If the x_i^R are random variables with mean $E(x_i^R) = \mu_i$ and if $E(x_i^R \varepsilon_j) = 0$ and $\text{Cov}(x_i^R, x_j^R) = 0$ for $i \neq j$, it happens that

$$\sigma_y^2 = \sigma_{yR}^2 + \sigma_{y\varepsilon}^2 = \sum_{i=1}^{m} \sigma_{x_i}^2 f_i^2 + \sum_{i=1}^{m} \sigma_{\varepsilon,i}^2 f_i^2$$

Second, if N independent and simultaneous measurements are made of each x_i^R, it is possible to decouple the random process noise from the error attributable only to the measurement, which can be estimated by

$$S_{\varepsilon i}^2 = \frac{1}{N-1} \sum_{j=1}^{N} (x_{ij}^0 - \bar{x}_i^0)^2$$

To illustrate the use of this first order approximation, return to the example of particulate measurement using EPA method 5. Brenchley and coauthors (1973) provide most of the data necessary for such an illustration, and what are not provided can be inferred. Taking as an example a new coal-fired power plant with rate of heat input of 174 $\times 10^6$ BTU/hr. and assuming that the measurement of the hourly particulate emission rate shows a level consistent with a standard of 0.10 lbs./10^6 BTU, the test may be scaled to produce an emission rate of 17.4 lbs./hr. It turns out that using representative data on individual measurement errors the overall variance due solely to measurement error, to a first order approximation, is quite large relative to the estimated emission rate. Thus the standard deviation of PMR measurements is approximately 2.5 lbs./hr. The 68 percent confidence interval for this measurement is therefore from 14.9 to 19.9 lbs. per hr., and the 95 percent confidence bounds are 12.4 and 22.4 lbs. per hr.

Thus, measurement error can be seen to have a potentially large impact on the ability of the agency to draw inferences about discharger compliance. It should be emphasized, however, that the above illustration concentrates on that aspect of the dual problem, ignoring the matter of process (discharge) variation over time. Overall uncertainty, as represented by the overall approximation to the variance, may reflect in any particular case either process variation or measurement error predominantly, or both may be of about the same importance.

The next section deals with the all-important matter of drawing inferences from data of the sort just discussed.

Statistical Errors and the Power of a Statistical Test

The origin of random and systematic errors in the monitoring and enforcement problem now has been discussed. Other matters discussed include how such errors affect the simplest direct pollution discharge measurements, both instantaneously and over averaging times; how to correct for systematic error in measurement; and how it is possible to approximate the overall effect of multiple random error sources from the multiple measurements actually necessary to the estimation of discharges. The introduction to the chapter, however, stressed that the effect of these errors was to introduce uncertainty about the true status of each discharger being monitored for compliance with a standard. This uncertainty can take two forms: the possibility that an apparent violation is not a true violation or that a true violation shows up as apparent compliance. By carefully defining the situation, it is possible to measure these alternatives. This quantification of error types is central to understanding the full difficulty faced by the regulatory agency in monitoring and enforcing compliance with its regulations.

Consider the problem of sampling from a series of randomly distributed "product pieces" or "discharges" with the aim of determining whether the mean product quality or the mean discharge level is μ_a, the null hypothesis; or whether the process in fact has slipped out of control so that the discharges are now being drawn from a distribution with mean $\mu_r = \mu_a \pm \delta\sigma_D$ where σ_D^2 is the variance common to both distributions, and μ_r is the alternative hypothesis.[10] In the general discharge monitoring case the results of the last section can be used to observe that σ_D^2 is composed of the error due to random variation in the true discharges (σ_{yR}^2) and the error due to imprecision in the measurement instruments ($\sigma_{y\varepsilon}^2$). These components, in turn, may be approximated on the basis of knowledge of the functions relating actual measurements to estimated discharges.

But for simplicity of exposition here, it will be convenient to revert to notation referring only to a measured quantity, x, as when the pollutant of interest can be measured directly. The variance of the distribution of draws of x is σ_x^2, invariant to assumed displacements of the mean, μ, and understood to be composed of variance in the x values themselves and imprecision in their measurement. Assume that individual observations of x are made in groups (samples) of n. Using these observations, it is possible to estimate the mean and standard deviation of x by the sample mean $\bar{x}$ and the sample standard deviation s_x. With these statistics one tries to determine whether the process is such that μ_a is the true mean of the stream of x values, or whether the process mean has been shifted—a violation is occurring—such that the true mean is $\mu_r = \mu_a + \delta\sigma_{\bar{x}}$.

---

[10] For discharges, one might well be interested only in the alternative hypothesis $\mu_r = \mu_a + \delta\sigma_{\bar{x}}$, since over-control is not a problem for the enforcement agency.

In this problem context there are three statistical measures of special importance.

- Type I error probability, often referred to as α: probability of finding a false violation—identifying μ_r as the mean when the true mean remains at μ_a. (In discharge monitoring, a type I error may be thought of as a false positive.)
- Type II error probability, often referred to as β: probability of missing a true violation—identifying μ_a as the true mean when the true mean has moved to μ_r. (A type II error is a false negative.)
- Power of test, $(1 - \beta)$: probability of not missing a true violation—identifying μ_r as the true mean when that is in fact the case.

The probability of type I error, α, is the probability of rejecting the null hypothesis, $\mu = \mu_a$, when no violation has occurred (that is, accepting the alternative hypothesis of a violation when the null hypothesis is in fact true). The probability of type II error, β, is the probability of accepting the null hypothesis that no violation has occurred (that is, rejecting the alternative hypothesis $\mu = \mu_r$ when the alternative hypothesis is true).

These error probabilities are interrelated. As the acceptable level of type I error probability is increased, other things equal, the power of the test increases. They are also functions of choices made about sample size, n, and about the level of alternative hypothesis, μ_r. As sample size increases, the power of the test based on the sample increases. And, other things equal, as the difference between the null and alternative hypothesis increases (as $\mu_r - \mu_a$ increases), the power of the test being applied increases.

In social science, there is seldom choice about sample size, and normally a simple alternative hypothesis, $\mu_r = \mu_a + \delta\sigma_{\bar{x}}$ where μ_r is known, cannot be formulated. Instead the researcher or agency is presented with a sample whose size is not a decision variable. However, in engineering applications such as those involved in monitoring it is often convenient to couch the problem in terms of solving for sample size n, given α, β and a specified μ_r that may represent, for example, the only possibility for machine maladjustment or the quality level beyond which customers will not accept the product (Blake, 1979, pp. 390–392). This kind of exercise is at the heart of the technique of statistical quality control, which is applied in chapter 6.

Suppose it is desired to test H_o: $\mu = \mu_a$ against H_1: $\mu = \mu_r \neq \mu_a$, where σ is known. This is a two-sided test of a simple null hypothesis against a composite alternative. The test rejects H_o if $|\bar{x} - \mu_a| > k$ where

k is a constant delimiting the boundary of the critical region of the test. The constant k is determined by the required probability α of a type I error, the probability of an observation falling outside the control limits when the process is in control. Formally, it is possible to write (Guttman, Wilks, and Hunter, 1971, p. 241):

$$\alpha \equiv P(|\bar{x} - \mu_a| > k \mid \mu = \mu_a) \qquad (9)$$

or

$$\alpha \equiv 1 - P(\mu_a - k < \bar{x} < \mu_a + k \mid \mu = \mu_a) \qquad (9a)$$

where $P(x|y)$ is to be read as the probability of x, given (or conditional on) y, and $|\cdot|$ denotes absolute value.

The probability of type II error, β, can be defined only in terms of a simple alternative hypothesis $\mu = \mu_a + \delta\sigma_{\bar{x}} = \mu_r$, where the critical value k must be chosen to be consistent with the probability of type I error. Given μ_r and k, the value of β depends on sample size, n. Type II error probability is defined as:

$$\beta = P(\mu_a - k < \bar{x} < \mu_a + k \mid \mu = \mu_r) \qquad (10)$$

A clearer understanding of the notion of type II error and its relation to type I error may be produced by a graphic demonstration (for an algebraic derivation, see Guttman, Wilks, and Hunter, 1971, pp. 155–161, p. 242). Figure 5-4 shows a one-tailed test with significance level α. The normal distribution of a random variable with a mean of μ_a in the unstandardized form and a mean of 0 in the standardized form is illustrated in panel I. The transformation required to move from the unstandardized to standardized form (that is, subtraction of μ_a and division by σ_x to produce a standard normal variate with mean 0 and standard deviation 1) is also shown. The critical value, k, in the unstandardized case is $\mu_a + z_\alpha\sigma_{\bar{x}}$ with significance level α. In panel II the alternative hypothesis $\mu = \mu_r$ is posited, and the variables are standardized at μ_r rather than μ_a. If we define δ to be $(\mu_r - \mu_a)/\sigma_{\bar{x}}$, then $\mu_r = \mu_a + \delta\sigma_{\bar{x}}$, and the 0 point in panel I becomes $-\delta$ in panel II. In panel III the standard normal distributions from panels I and II are overlayed on each other so that, given z_α, the type II error probability is given by the stippled area.

From the graphic presentation it should be clear that as the separation between null and alternative hypotheses increases, β falls for constant α (and vice versa). A similar effect is achieved by decreasing the standard

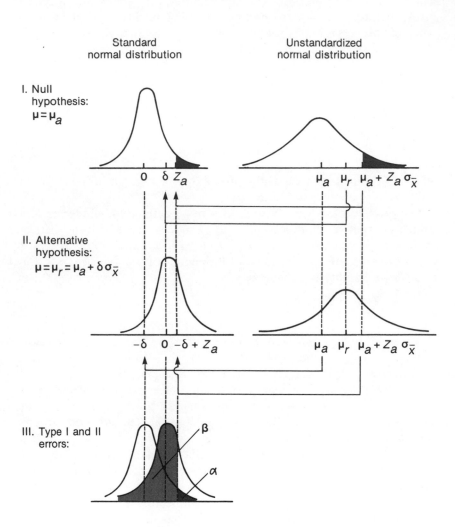

Figure 5-4.   Graphic demonstration of the error probabilities, one-tailed test.   *Note*: To convert an unstandardized normal variate x with mean μ_x and variance σ_x^2 to a standardized normal variate z, the transformation is

$$z = \frac{x - \mu_x}{\sigma_x}.$$

deviation. And, as already noted, for constant separation, an increase in α implies a decrease in β (and vice versa).

The implications of these error probabilities for the design of monitoring schemes will be worked out in chapter 6, where quality control techniques are applied under the assumptions that cheating is not an option and that the benefits of control are known; and in chapter 7, when those restrictions are relaxed. But to illustrate immediately the relation among sample size, difference between null and alternative hypotheses, and error probabilities, table 5-1 is provided. In this table, σ is kept constant at 1.0, and α is required to be 0.05 (so that $z_\alpha = 1.96$). The single-tailed possibilities, with μ_r larger than μ_a by various multiples of σ, are examined.

The following effects are illustrated in the table:

- For fixed α and sample size, increasing the spread between null and alternative hypothesis *relative to* σ increases the power of the test. Thus, the power of a given monitoring method can be increased by improving the precision (lowering the $\sigma_{y_e}^2$) of the instrumentation. The same instrumentation will be more powerful when applied to a discharge stream with smaller variability (σ_{yR}^2).

- For fixed α and spread between hypotheses, increasing the sample size increases the power of the test dramatically. This occurs because of the effect of increasing sample size on the estimate of the sample variance σ_x, an effect noted earlier in the chapter when sample statistics were described.

Conclusion

This chapter has included a discussion of how randomness enters the monitoring and enforcement problem—through random elements in source discharges and through random errors in measurements of those discharges. Some basic statistical formulas were set out to show how those random elements are transformed by the summarization of repeated measurements. An approved technique for pollution discharge monitoring—EPA's method 5 for measuring particulate mass emission rate—was used to illustrate the complexity of determining overall measurement error in realistic cases, and the overall error bounds for the method were calculated from example data. The existence of random error was next shown to lead to probabilities of errors of *inference*— that is, to arriving at the wrong conclusions on the basis of measured values. Two varieties of such inferential error were identified and de-

Table 5-1. Sample Calculations of Type II Error Probability for Different Sample Sizes Given a Constant Type I Error Probability

$\mu_a - \mu_r$	$\dfrac{\mu_a - \mu_r}{\sigma_{\bar{x}}}\, z_\alpha$			β			$1 - \beta$		
	$n=1$	$n=2$	$n=3$	$n=1$	$n=2$	$n=3$	$n=1$	$n=2$	$n=3$
0	1.96	1.96	1.96	.9750	.9750	.9750	.0250	.0250	.0250
−.05	1.91	1.8893	1.8734	.9719	.9706	.9695	.0281	.0294	.0305
−.25	1.71	1.6064	1.5270	.9564	.9459	.9366	.0436	.0541	.0634
−.50	1.46	1.2529	1.0940	.9279	.8949	.8630	.0721	.1051	.1370
−.75	1.21	.8993	.6610	.8869	.8158	.7457	.1131	.1842	.2543
−1.00	.96	.5458	.2279	.8315	.7074	.5902	.1685	.2926	.4098
−1.25	.71	.1922	−.2051	.7611	.5762	.4188	.2389	.4238	.5812
−1.50	.46	−.1613	−.6381	.6772	.4359	.2617	.3228	.5641	.7383
−1.75	.21	−.5149	−1.0711	.5832	.3033	.1421	.4168	.6967	.8579
−2.00	−.04	−.8684	−1.5041	.4840	.1926	.0663	.5160	.8074	.9337
−2.25	−.29	−1.222	−1.9371	.3859	.1109	.0264	.5141	.8891	.9736
−2.50	−.54	−1.5755	−2.3701	.2946	.0576	.0089	.7054	.9424	.9911
−2.75	−.79	−1.9291	−2.8031	.2148	.0269	.0025	.7852	.9731	.9975
−3.00	−1.04	−2.2826	−3.2362	.1492	.0112	.0006	.8508	.9888	.9994
−3.25	−1.29	−2.6362	−3.6692	.0985	.0042	.0001	.9015	.9958	.9999

Note: $\alpha = 0.05$ so that $z_\alpha = 1.96$ for a one-tailed test. $\sigma = 1.0$.

fined: false positives (probability denoted by α) and false negatives (probability denoted by β). The probabilities of the two errors are related and, indeed, other parts of the situation remaining the same, must move in opposite directions.

There are three very important messages here for the design of monitoring and enforcement schemes. First, in the practical world of the regulator, errors of inference do exist and cannot be ignored or assumed away. Second, the achievable error probabilities depend, among other things, on the monitoring technology *and* the number of samples per test measured with that technology. Third, and perhaps most important of all, error probabilities do not exist in a vacuum, nor are they ordained by any supernatural force; they must be *chosen by the decision maker*. Thus, whether a monitoring system is shaded toward protecting sources against false notices of violation or toward protecting the environment against undiscovered violations of standards is a policy decision. Furthermore, it is an unavoidable decision with substantial importance for subsequent results. One might fairly argue that it is a decision that should be made publicly and explicitly. Currently, such a decision is made through a series of apparently unconnected decisions on definitions of standards, and on allowances for excessive emissions both in short-run quantity and in timing.

APPENDIX 5-A

Measurements and Calculations for the Estimation of Particulate Discharges Using EPA Method 5

As noted in the text, the particulate mass emission rate (PMR) in pounds per hour can be thought of as the product of two subsidiary quantities, Q and C, where Q is stack gas flow rate and C is the concentration of particulates in the stack gas. So PMR $= Q \cdot C$.

C is the slightly less complicated of these two quantities. It is obtained from:

$$C = \frac{m}{V_{ms}} \tag{A-1}$$

where m is the pounds of particulate material collected in the sample device from the sample volume of gas and V_{ms} is the volume of dry gas sampled, at standard conditions (which is to say, at a temperature of 530° absolute and 29.92 inches of mercury).

V_{ms} in turn is calculated from the measurements made at the dry gas meter.

$$V_{ms} = k_1 V_m \frac{P_b + \Delta m}{T_m} \qquad \text{(A-2)}$$

where P_b is the barometric pressure at the orifice meter (the end of the sampling train)

      Δm is the average pressure drop across the orifice meter (measured in same units as P_b)

      V_m is the volume of gas actually measured by the dry gas meter

      k_1 is a constant relating to standard conditions of temperature and pressure

Also, $T_m = \frac{1}{2}(T_{m1} + T_{m2})$, and T_{m1} and T_{m2} are meter inlet and outlet gas temperatures, respectively.

Calculating Q is somewhat more complicated, involving the same kinds of corrections plus others for water vapor and for the nature of the instrument itself. Thus, to begin with, Q is given by:

$$Q = 3{,}600\,(1 - B_{\text{H}_2\text{O}})(V_s)Ak_1 \frac{P_s}{T_s} \qquad \text{(A-3)}$$

3,600 is the "correction" from the velocity in cu. ft./sec. as measured to cu. ft./hr. as desired. (But note that the actual measurement of both volume and particulate matter actually is done over some period t, so that *total* particulates and *average* velocity are measured.)

k_1 is as above

A is the cross sectional area of the stack being sampled

P_s is actual static pressure in the stack

T_s is actual (absolute) temperature in the stack

$B_{\text{H}_2\text{O}}$ is the fraction of the stack gas that is water vapor

V_s is the instantaneous velocity of the stack gas.

Both B_{H_2O} and V_s are themselves determined from other formulas and measurements. Thus,

$$B_{H_2O} = \frac{V_{ws}}{V_{ms} + V_{ws}} \qquad (A\text{-}4)$$

where V_{ms} is the volume of dry gas sampled as determined above and V_{ws}, the volume of water vapor in the stack gas, comes from:

$$V_{ws} = k_2 V_h$$

where k_2 is a correction factor involving density and molecular weight of water, the vapor pressure of water as steam at standard conditions, as well as the standard temperature and pressure correction; and V_h is the volume of water trapped in the impinger section of the sampling train.

Finally, V_s is obtained from the pressure drop across the "pitot-tube"— a velocity measuring device—suitably corrected.

$$V_s = k_3 (\Delta P)^{1/2} \left[\frac{T_s}{P_s M_s} \right]^{1/2} \qquad (A\text{-}5)$$

where k_3 corrects for the specific pitot-tube method shown in figure 5-3; ΔP is the measured difference between the pressure on the pitot-tube orifice facing the gas flow and that facing away from the flow; T_s and P_s are as defined above; and M_s is the molecular weight of the stack gas, which is itself calculated from separate estimation of the constituents of the dry gas (CO_2, N_2, and the like) and the previously calculated B_{H_2O}, or water vapor fraction. Thus, M_s is given by

$$M_s = (M_d)(1 - B_{H_2O}) + (18)B_{H_2O} \qquad (A\text{-}6)$$

and

$$M_d = B_{CO_2}(44) + B_{CO}(28) + B_{N_2}(28) + B_{O_2}(32) \equiv \sum B_i M_i \qquad (A\text{-}7)$$

The numbers in parentheses are the applicable molecular weights, referred to later as M_i, and B_i are the measured fractions of the stack gas accounted for by its major constituents. (The fractions accounted for by SO_2, NO_x, and other pollutants are ignored in this approximation.)

When these formulas are combined into one expression giving PMR as a function of known constants and measured quantities, a very complicated equation is obtained. It is not reproduced here.

References

Beavis, Brian, and Martin Walker. 1983. "Achieving Environmental Standards with Stochastic Discharges," *Journal of Environmental Economics and Management* vol. 10, no. 2 (June) pp. 103–111.

Berthouex, P. M., W. G. Hunter, L. Pallesen, and C. Y. Shih. 1976. "The Use of Stochastic Models in the Interpretation of Historical Data from Sewage Treatment Plants," *Water Research* vol. 10, pp. 691–692.

Bevington, Philip R. 1969. *Data Reduction and Error Analysis for the Physical Sciences* (New York, McGraw-Hill).

Blake, I. F. 1979. *An Introduction to Applied Probability* (New York, John Wiley and Sons).

Brenchley, D. L., C. D. Turley, and R. F. Yarmac. 1973. *Industrial Source Sampling* (Ann Arbor, Michigan, Ann Arbor Science Publishers).

Cochran, William G. 1977. *Sampling Techniques*, 3d ed. (New York, John Wiley & Sons), chapter 1.

Davidson, Jill E., and Phillip K. Hopke. 1984. "Implications of Incomplete Sampling on a Statistical Form of the Ambient Air Quality Standard for Particulate Matter," *Environmental Science and Technology* vol. 18, no. 8, pp. 571–580.

Environmental Science and Technology. 1978. "Federal Environmental Monitoring: Will the Bubble Burst?" (November) pp. 1264–1269.

Evans, John S., Douglas W. Cooper, and Patrick Kinney, no date. "On the Propagation of Error in Air Pollution Measurements." Unpublished paper (Boston, Massachusetts, Department of Environmental Science and Physiology, Harvard University, School of Public Health).

Gardenier, Turkan K. 1982. "Moving Averages for Environmental Standards," *Simulation* vol. 39, no. 2 (August) pp. 49–58.

Guttman, Irwin, S. S. Wilks, and J. Stuart Hunter. 1971. *Introductory Engineering Statistics* (New York, John Wiley & Sons), chapters 10 and 11.

Hunter, J. Stuart. 1980. "The National System of Scientific Measurement," *Science* vol. 210, no. 21 (November) pp. 869–874.

Larsen, Richard J., and Morris L. Marx. 1981. *An Introduction to Mathematical Statistics and Its Applications* (Englewood Cliffs, New Jersey, Prentice-Hall).

National Academy of Sciences. 1977. *Environmental Monitoring, A Report to the U.S. Environmental Protection Agency from the Study Group on Environmental Monitoring* (Washington, D.C., National Academy of Sciences, National Research Council).

Parzen, Emanuel. 1960. *Modern Probability Theory and Its Application* (New York, John Wiley & Sons).

Shaw, R. W., Jr., M. V. Smith, and R. J. Paur. 1984. "The Effect of Sampling Frequency on Aerosol Mean Values," *Journal of the Air Pollution Control Association* vol. 34, no. 8, pp. 839–841.

6
A Statistical Quality Control Model

Understanding of the monitoring problem can be improved by structuring the analysis in a way that directly incorporates the stochastic nature of actual discharges and the measurement errors of monitoring instruments. These complications suggest that the quality control literature, originating in the needs of manufacturers to ensure that their products meet certain quality specifications, might hold some useful lessons.[1] This chapter summarizes rather briefly what can be learned from this literature—both by way of answers and, more importantly, by way of further difficulties and questions. The material is based on Vaughan and Russell, 1983.

Definition of the Problem

Consider a single point source of pollution, be it a firm or a municipality, that can pick a target discharge per unit time, $\hat{D}_t$. Actual discharges in any particular unit of time are drawn independently from a normal

---

[1] This book is by no means the first to see this possible parallel. Berthouex and Hunter (1975) and Berthouex, Hunter, and Pallesen (1978), for example, applied quality control methods to monitoring sewage treatment plant performance, adopting the point of view of the plant management.

distribution,[2] with mean μ_D and variance σ_D^2. For convenience assume further that σ_D^2 is independent of the chosen target; and that unless otherwise stated, $\hat{D}_t$ is a constant target for $t = 1, \ldots, T$, which may be specified by the regulation. As a matter of substance assume that the owner/operator will not try to cheat; that is, if the responsible agency announces a standard, the discharger will pick a $\hat{D}_t$ designed to meet it, although violations may occur for reasons beyond his control. This is consistent with the voluntary compliance model of chapter 4 and is critiqued later in this chapter. A final model, in which cheating is an option, is offered in chapter 7.

Denote the standard by μ_a and assume that this is defined to be the required *expected value* of discharges. Note that this approach to standard setting begs the question of how standards actually are defined, a matter discussed in chapter 3. Later in this chapter this matter is taken up again briefly.

In order to check on the compliance of the discharger with the standard the agency is assumed to be able to observe actual discharges, without bias but with random error. Only one method of observation, with a cost function to be described below and with precision of σ_I^2, is available for these observations. The agency is free to take up to n observations simultaneously, and these constitute a sample. It can choose the (fixed) time interval between samples, s. The reading of the method, X_i for a single observation, is assumed to be in quantity per unit time, directly comparable to D_t, to avoid worrying about separate measurements of flow and concentration. A sample mean will be denoted $\overline{X}$ for size n. The resulting variance due to measurement error will be σ_I^2/n. For the agency's purposes, the total variance of the "system," reflecting discharge and measurement variance will be $\sigma_T^2/n = (\sigma_D^2 + \sigma_I^2)/n$, assuming independence of measurement and process.

The agency's problem, which will be addressed through a quality control approach, is to design a monitoring scheme for discharges that is socially optimal in the sense that it minimizes the sum of social damages

---

[2] Most effluent data are not normally distributed, but they often can be made so by a suitable transformation. The independence assumption means that the observations are not serially correlated and is made here for convenience. It may often be violated in the real world. For a theoretical discussion of the consequences of violation of this standard assumption on the performance of the detection techniques, see Goldsmith and Whitfield (1961) and Bagshaw and Johnson (1975). For an application of quality control methods to data from a sewage treatment plant where serial correlation is present, see Berthouex, Hunter, and Pallesen (1978). For a clever treatment of some control chart techniques to detect lack of control in the presence of autocorrelation, see Gardenier (1982).

and costs attributable to the scheme.[3] These damages and costs are comprised of:

- The net social benefits lost when discharges violate the standard. Because of the statistical setting created by the assumptions, this can be thought of as the damages from type II error—undiscovered violations.
- The cost of running the monitoring scheme itself.
- The cost of looking for and correcting the cause of a true violation.
- The cost of (fruitlessly) looking for the cause of a reported violation that is not a real violation. This is taken to be the cost of type I error—false positives.

One way to approach this problem is through expected values of the stochastic quantities of interest. To make this possible, consider an operational cycle, each repetition of which would differ in length but which, through endless repetition, would define the expected values of the various cycle parts (See, for example, Duncan, 1956). This cycle is depicted in figure 6-1. At the beginning of the cycle, discharges are in control, so that their expected value is μ_a, the standard. After some time T_a, there is a "failure" resulting in loss of discharge control, and the expected value of discharges leaps to μ_r. After some lag, the agency discovers the change in expected values and instructs the discharger to find and correct the cause, after which the process returns to the in-control state, and the cycle repeats. No fines or punishments are levied for the discovered failure. The decay in reliability of the overall process generating discharges is assumed to be captured by a negative exponential density function, such that the expected time to failure $E(T_a)$ is $1/\lambda$, where λ is often called the hazard rate (Duncan, 1956, p. 299). Again this is consistent with the model of voluntary compliance.

The loss of net benefits per unit time during the out-of-control period will be written as $B(\mu_a) - B(\mu_r) \equiv B_a - B_r$, where it is understood that expectations are involved. Finally, note that though false alarms are not shown in figure 6-1, they can be expected to occur during any period of monitored, in-control operation.

In this context, a specific monitoring scheme has three essential characteristics: the sample size taken, n; the interval between samples, s; and the rule triggering the violation alarm. These characteristics are in

---

[3] Optimization models of process control are discussed and critically reviewed by Chiu and Wetherill (1973) and Gibra (1975).

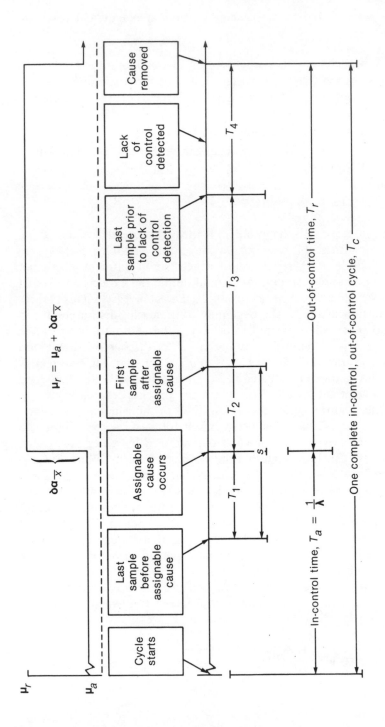

Figure 6-1.   Schematic of a quality control cycle.   *Source:* Adapted from A. L. Goel and S. M. Wu, "Economically Optimum Design of CUSUM Charts," *Management Science* vol. 19, no. 11 (July 1973) pp. 1271–1281.

turn linked to the minimization problem through their influence on type I and II errors and thus on the key variables, called average run lengths for in- and out-of-control operation:

L_a, the expected number of samples from cycle beginning to a false alarm;

L_r, the expected number of samples from a control failure to its discovery.

Derivation of the Function to Be Minimized

The objective function to be minimized is the sum of the benefits lost when the process operates in violation of the standard and the costs of running the monitoring operation, including the costs of finding the causes for and correcting real violations and of trying to deal with false alarms. For our purposes, these will be expressed per unit of time. A compact expression for the objective function is:

$$£ = (B_a - B_r) \gamma + C_1$$

where γ is the expected fraction of time the process is out of control, B_a is the net benefit per unit time when the process is in control, and B_r is the net benefit when the process is out of control. C_1 is the cost per unit time of running and following up on the monitoring scheme. As a general matter, by increasing C_1 it is possible to decrease γ.

C_1 may be decomposed into three elements, all expressed per unit time (Wetherill, 1977).

C_2 = Cost of sampling and administration;
C_3 = Cost of looking for and correcting a failure when one exists;
C_4 = Cost of looking for failure when none exists.

These elements in turn may be defined as follows:

$$C_2 = \frac{b + cn}{s} \text{ is the usual expression for sampling cost, where} \tag{1}$$

$b \equiv$ constant per-sample cost;
$c \equiv$ cost per observation taken;
$n \equiv$ number of observations taken per sample;
$s \equiv$ sampling interval, units of time that elapse between taking samples of n articles.

The above expression says that there is a component of sampling cost, b, which is independent of sample size and avoidable if a sample is not taken; that is, b represents a variable cost with respect to the number of samples but a constant per sample.

More generally, the monitoring program will also involve fixed costs associated with different types of capital equipment. And each monitoring device may involve different values of b and c as well as different capital cost and different measurement error. Thus, a more general form for the monitoring cost function would be:

$$C_2^i = \frac{K^i \cdot crf}{u} + \frac{b^i + c^i n}{s} \tag{2}$$

where $K^i \equiv$ total capital cost of equipment type i
    $u \equiv$ time units per year
    $s \equiv$ sample interval
    $crf \equiv$ appropriate capital recovery factor

In what follows, it will be assumed that monitoring equipment of a particular type is already in place, and the short-run form of the monitoring cost function will be used. The fuller choice model is much more complicated and perhaps more realistic, but it neither avoids nor better illustrates the central issues.

The final two cost components are, conceptually at least, straightforward:

$C_3 = g(n,s)W$ is the cost of finding causes of and correcting real failures; where:

    $g(n,s) \equiv$ average number of occasions per unit time that the process goes out of control;

    $W \equiv$ cost of looking for an assignable cause of failure or violation when it exists.

$C_4 = \ell(n,s)Y$ is the cost of searching for the causes of false alarms; where:

    $\ell(n,s) \equiv$ average number of false alarms per unit time;

    $Y \equiv$ cost of looking for a false alarm

C_3 and C_4 are both real costs for society even though they may be borne by the sources being monitored, when they are forced to react to reports of violations, some of which are real and some false.

Both g and ℓ are functions of sample size and interval. This may seem counter intuitive in the case of g, the number of losses of control per unit of time, for this would seem to depend only on the characteristics of the failure-producing process and to be independent of the sample scheme. The trick is that the overall length of the process cycle is defined by the occurrence of one failure and its correction. The amount of time required for this to happen depends on the sampling interval and the precision of the control tests, the latter being functions of n, the sample size. Another way of saying this is to relabel $g(n,s)$ as the average number of *discovered* and corrected violations per unit time.

Thus, the full damage and cost function (called the "loss cost function" in the quality control literature) may be written as:

$$£ = \gamma(n,s)[B_a(\mu_a) - B_r(\mu_r)] + g(n,s)W + \ell(n,s)Y + \frac{b + cn}{s} \qquad (3)$$

This expression cannot be worked with directly, but must be translated into expected value terms via the control cycle and average run length concepts. The algebra necessary to do this is somewhat tedious and will not be discussed in the text, though the interested reader may find it in appendix 6-A. For purposes of understanding the application of quality control methods to monitoring program design, it will be sufficient to note that the resulting objective function expression depends in complicated ways on the sample size and sampling interval choice variables (n and s); on the assumed alternative hypothesis, or alternative discharge level (μ_r); and on the average run lengths (L_a, L_r), which are themselves in turn principally determined by the choices made for the error probabilities α and β. (Recall that for a given choice of α, the size of β depends on the separation of the hypotheses ($\mu_a - \mu_r$) and on the precision of measurement, which is a function of n.) In addition, as was explained at the beginning of the exposition, the objective function reflects the differential benefits between the in-control and out-of-control states. (These might be termed the *damages* of a violation.)

The particular relationships between policy choices and run lengths, and hence the choices of n and s that minimize the objective function, differ with the choice of detection rule. In the next sections, several common rules are discussed.

The Shewhart Method: Error Probabilities and Run Lengths

In this method of violation (failure) detection, each sample of size n is considered separately as it is drawn, and the information obtained from the sample is then effectively discarded. The "consideration" consists of comparing the sample mean (the estimate of actual discharge) to an upper control limit (UCL).[4]

$$UCL = \mu_a + Z_\alpha \frac{\sigma_T}{\sqrt{n}} \tag{4}$$

where Z_α is the value of the standard normal variate corresponding to chosen type I error, α. (See chapter 5 on error probabilities.) In this scheme, the expected value of in-control run length (the expected number of samples until a false alarm) is especially simple: $L_a = 1/\alpha$.

As described in chapter 5, type II error probability, denoted β, and its complement, the "power of the test," $1 - \beta$, are determined by the chosen α, sample size n, and the alternative hypothesis μ_r (the out-of-control expected value of discharges). In fact, β may be calculated, from the relation:

$$\beta = \Phi[Z_a + \sqrt{n} \, (\mu_a - \mu_r)/\sigma_T] \tag{5}$$

where $\Phi[\ ]$ indicates the value of the cumulative normal at the upper given limit of integration, the lower limit being $-\infty$. L_r, in turn, is simply $1/(1 - \beta)$. Notice that in this application $\mu_r > \mu_a$ by assumption, so that the greater the difference between the standard and the hypothesized rate of discharge in failure conditions, the more negative the upper limit of integration, the smaller β, and the lower L_r.

Solution of the optimal monitoring scheme problem then can be thought of as follows:

Given the cost parameters b,c,W,Y; the failure-correction parameters D and τ; the "hazard rate" λ; and net benefits $B_a(\mu_a)$ and $B_r(\mu_r)$:

1. Assume μ_r is also "known"; that is, for example, assume that when failure occurs, it is always complete failure of the treatment equipment, so that discharges either have expected value D_t, or $[1/(1 - d)] \, D_t$ where d is the removal efficiency of the equipment in place. Then $B_a(\mu_a) - B_r(\mu_r)$ is also known.

    a. Pick α_1, trial type I error, and calculate L_a and Z_α.
    b. Choose s_1, trial sample interval; and n_1, trial sample size. Solve for β and L_r. Calculate the value of £.
    c. For the same α_1, s_1 pair, try new sample sizes n_i for $i = 2, \ldots, I$.

---

[4] This discussion proceeds for a one-tailed test because in most pollution control situations interest is in discharges in excess of the standard only.

    d. For the same α_1, pick new sample intervals s_j and repeat steps b and c $(j = 1, \ldots, J)$.

    e. Pick new type I error probabilities α_k and repeat steps b, c, d for $k = 1, \ldots, K$.

Choose the combination α_k, n_i, s_j that makes £ a minimum. Then, to put the scheme in operation, calculate $Z_{\alpha_k}\sigma_T/\sqrt{n_i}$ and match each sample mean, $\overline{X}$, against $\mu_a + Z_{\alpha_k}\sigma_T/\sqrt{n_i}$. More efficient search techniques may be employed, and, in the case of the Shewhart detection scheme, it is even possible to proceed analytically; that is, the first partials $\partial£/\partial\alpha, \partial£/\partial n$, and $\partial£/\partial s$ can be written down, although the algebra is extremely tedious, and numerical methods would have to be used to solve the resulting set of simultaneous nonlinear equations.[5]

2. When μ_r is not "known" in advance, the problem is considerably more difficult. The task is not simply to specify an optimal μ_r along with α, n, and s, for the actually realized μ_r will not be under the agency's control. Indeed, the realized μ_r will not really be under anyone's control if breakdowns of different degrees of seriousness in different parts of the production and treatment chain are possible, as might well be expected. Rather the problem is to pick the best μ_r for the monitoring scheme given that a different out-of-control level of discharge in general will be realized. We can write this problem as:

$$\min_{(\hat{\mu}_r)} \sum_{\mu_r = \mu_a + 1}^{\theta} £[\alpha(\hat{\mu}_r),\quad n(\hat{\mu}_r),\quad s(\hat{\mu}_r),\quad \mu_r]$$

where $\hat{\mu}_r$ is the hypothesized out-of-control discharge rate and μ_r is the realized rate. θ could be the largest imaginable discharge from the particular plant.[6]

Other Detection Methods

The problem of minimizing the sum of costs and damages can be approached with other, more sophisticated detection methods as well. One possibility is CUSUM, a simple transformation of the Wald sequential

---

[5] One might guess that local optima are the rule rather than the exception in this problem, so alternative starting points ought to be tried. For an exposition of this solution method, see Goel and Wu (1973).

[6] Treatments of this problem in the literature do not generally recognize the dependence of B_r on the realized μ_r.

hypothesis testing technique. (On the technique, see Ewan, 1963; Guttman, Wilks, and Hunter, 1971. For its application in optimization, see Goel and Wu, 1973.) When using CUSUM, the difference between a sample value $\overline{X}$ and the in-control expected value μ_a is not, as in the Shewhart method, compared by itself to $Z_\alpha \delta_T / \sqrt{n}$. Rather, the values of $\overline{X}_j - \mu_a$ are continuously summed as sampling proceeds over the index j. The type I and II errors here depend on ratios of products of density function values (a likelihood ratio test). The numerator consists of the product of density function values for the sample observations minus μ_a; given that the alternative hypothesis is correct (that the density function is centered on μ_r). The denominator is the product of density function values for the sample observations minus μ_a; given that the density function is centered on μ_a.

For the CUSUM test, determination of run lengths for choices of μ_r and n is not a matter of solving simple expressions involving normal distribution functions. While approximation formulas do exist (Goldsmith and Whitfield, 1961; Reynolds, 1975), these are close approximations only in small neighborhoods. For exact calculation, nomograms based on sophisticated numerical methods often have been used in the past (for example, in Kemp, 1961, Wetherill, 1977, and Goel and Wu, 1971).

A third possible method for manipulating sample observations and signaling violation is the MOSUM, which may be thought of as a standardized and truncated CUSUM; that is, the sums of differences $\overline{X}_j - \mu_a$ are summed only for a predetermined number of samples, G. As each sample point is added, a previous one is dropped from the moving sum. In addition, the sum is divided by the process standard deviation. Again, working with this detection method in the optimization framework is more difficult because closed forms do not exist for L_r as a function of the choice variables[7] (Bauer and Hackl, 1978, 1980). But, rather than dwelling on specific detection methods and their complications, it is now necessary to turn to the most serious questions raised by the general quality control approach just briefly described.

The Quality Control Approach to Monitoring: Questions

Using quality control methods at least allows a monitoring agency to come to grips directly with the unavoidable facts of stochastic discharges

---

[7] Less importantly, calculations involving type I error, α, also require approximations because of nonzero covariances between MOSUMs.

and measurement imprecision. Furthermore, it appears to allow the design of optimal sampling schemes, a feature that always appeals to economists. But to get this far with the quality control approach it is still necessary to make some heroic assumptions. The utility of the approach will depend on how hard or easy it is to relax those assumptions. Said in a slightly different way, the prospective utility of quality control methods in source discharge monitoring depends most importantly on the answers to three questions:

- Is the notion of a discharge standard as an expected value consistent with society's intention, and with the reality of laws and regulations?
- How difficult is it to produce the source-specific benefit functions necessary to specifying the net benefits of in-control and out-of-control operations?
- Is it possible to deal with the possibility of willful violation (cheating) within the quality control approach?

How Are Discharge Standards Defined?

This may seem a silly or redundant question, especially in view of the brief discussion of standard setting in chapter 3. But the quality control approach makes very clear the value of knowing more precisely what kind of standards are to be enforced. The rhetoric of U.S. environmental policy suggests that discharge standards are supposed to be upper limits, with no violations tolerated. This position, for example, is implicit in the arguments of those who espouse standards with stiff penalties for violation instead of effluent charges, which appear to leave the choice of discharge level entirely to the discharger (see survey results, Kelman, 1981). In this view, the place of statistical considerations would be confined to allowing for monitoring instrument error. Such an approach need not be a result of emotional responses only. It falls out of a model in which the legislature or agency has determined that very large losses might result from discharges above the chosen limit.

In contrast, the view of the standard as a target (expected value) is supported not only by the reality of variation in production and treatment processes but also by an assertion that for most sources and most pollutants, damage functions (if they were but known) would display no serious discontinuities in the neighborhoods of the socially optimal (or at least the chosen) levels of discharge. And, whatever the rhetoric of legislators, federal and especially state agencies are enforcing standards as targets, allowing equipment breakdown or human error to stand as a sufficient defense against a charge of violation. For evidence on this point, see chapters 2 and 3; Downing and Kimball, 1976; and Harrington, 1981. And for a description of the acceptable reasons for self-reported excess stack opacity under New Source Performance Standards, see Ruger, 1981.

Consider a restatement of the alternatives designed to make the implications for an enforcement agency more explicit. First, suppose the agency is to use a quality control criterion that signals that a process is out of control (that is, a standard has been violated) any time an outcome exceeds δ standard deviations from the design effluent concentration target of the control device, μ_a. This is a probabilistic monitoring criterion. Let s represent the published standard; μ_a the design effluent performance criterion of the quality control test (which approximately equals the average effluent performance of a large group of "ideal" plants); x_i the observation on the process outcome; and σ the standard deviation. In these circumstances, consistency in rulemaking and enforcement can be achieved only if the null hypothesis of the statistical quality control criterion is set properly. Following is an example, in terms of the Shewhart test, for samples of size one:

Consistent: Deterministic Rule, Probabilistic Enforcement

Rulemaking: $s = \mu_a$
Null hypothesis for monitoring: $x_i = s = \mu_a$
Violation signal: $x_i > \mu_a + \delta\sigma$ or $s + \delta\sigma$

Another possibility is to incorporate in the standard the uncertainty about the discharger's success in meeting the target. Thus:

Consistent: Probabilistic Rule, Deterministic Enforcement

Rulemaking: $s = \mu_a + \delta\sigma$
Null hypothesis for monitoring: not applicable—built into the rulemaking step
Violation signal: $x_i > s$

While consistent, this approach, by establishing once-and-for-all a particular degree of confidence for the monitoring scheme and making enforcement deterministic, seems to take discretion away from the enforcing authorities, who, in monitoring, might have preferred a degree of confidence at odds with that embodied in the regulation. That this is possible will be clear from the section on optimization above, since the optimal level of confidence is a function of reliability of the treatment equipment, sampling tests, and source-specific damages. However, as was described in chapter 3, this is in fact the common approach to standard-setting under U.S. air and water quality legislation.

Finally, observe that overcorrecting for the stochastic nature of the discharges being monitored will create an inconsistency that essentially works to the advantage of the discharger. Thus:

Inconsistent: Probabilistic Rule, Probabilistic Enforcement

Rulemaking: $s = \mu_a + \delta\sigma$

Null hypothesis for monitoring: $x_i = s$

Violation signal: $x_i > s + \delta\sigma = \mu_a + 2\delta\sigma$ $\qquad\qquad$ (6)

Enforcement of the published standard as if it were μ_a produces a $2\delta\sigma$ confidence interval. (For an example of this sort of confusion, see Rice, 1980, where the "standard" at issue is the required absence of some compound from a discharge stream, and a violation occurs when the compound is present.) Moreover, a probabilistic rule that allows the enforcement agency no latitude for probabilistic enforcement can always be manipulated into an equivalent deterministic rule with probabilistic enforcement by the agency if it knows the basis $(\mu_a, \delta\sigma)$ of the rulemaking decisions, where the δ for enforcement need not equal the δ for rulemaking used to establish $s = \mu_a + \delta\sigma$.

Can the Source-Specific Benefit Function Be Estimated?

There is no point in belaboring here the difficulties posed for the design of a discharge monitoring scheme by the necessity of knowing something about the losses society suffers when discharge standards are not complied with. Certainly progress is being made in the development of techniques for estimating pollution control benefits in amenity, health, aesthetic, and even "existence value" realms. (For a catalog of recent advances, see Kneese, 1984.) But much work remains to be done before either aggregate benefits for a particular policy decision over some geographically or politically defined area or damage estimates for individual sources of particular pollutants will be available. Indeed, the latter in principle may be impossible to develop where more than one source contributes to pollution, and nonlinearities characterize the meteorology, hydrology, or the dose-response relations. When regulations are written for individual stacks at large industrial facilities the practical problems are multiplied. This at least can be avoided by treating the plant as a unit (that is, by adopting the "bubble concept").

In the monitoring context there is yet one more difficulty associated with measuring $B_a - B_r$. Consider the two situations shown in figure 6-2. If *TB* represents the gross benefit of residuals discharge reduction

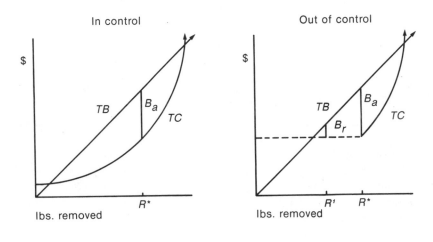

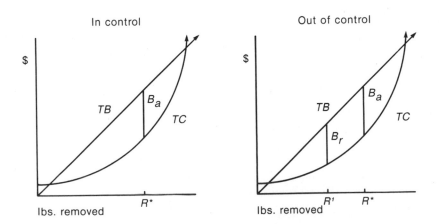

Figure 6-2.   Contrast in losses due to violations: malfunctions versus deliberate shutdown

and TC the total cost of treatment (exclusive of monitoring cost) and the standard is set optimally, net benefits when the process is in control, B_a, will be maximized at removal level R^*. Now, if the treatment equipment malfunctions in some way, so that removal falls to R^1 while total treatment costs remain fixed as in panel A, the value of B_r will be much smaller than it would be if the device operator could in some way simultaneously reduce removal efficiency and running cost as in panel B. Practically speaking, the enforcement agency can never know in advance which situation holds (or whether some intermediate possibility has occurred), so it can never precisely quantify $B_a - B_r$.

Can Quality Control Monitoring Work When Deliberate Cheating Is Possible?

The last observation in the section on determining $B_a - B_r$ implicitly raised the issue of deliberate violation. In crudest terms, as discussed in chapter 5, dischargers, knowing the agency to be operating a fixed sampling program appropriate to a situation in which random machine failure is the problem, might well adjust their discharges in a "battle-ment" pattern over time. Then there would be no control applied between sampling visits, but required controls would be operating during those visits.[8] Whether this would be profitable or even possible in any particular case would depend on the source's cost saving from disconnecting or shutting down its control device(s), the cost of the shutdown and startup, the period it would take for the equipment to "settle in" after startup, and the frequency of the announced sampling visits.[9] But whether such gross violation would be the rule or the exception, its specter should inspire thought about alternatives to the preannounced, fixed, sampling period s.

---

[8] Sample and visit are used interchangeably here. One might argue that self-monitoring, a common requirement, could still involve fixed, optimal sample intervals, for cheating, if it is to occur, would surely not depend on measurement timing when the polluter is measuring his own discharge. The timing of audits of self-monitoring performance may still be a problem, and here visits would equal samples.

[9] As discussed in chapter 3 there are further complications here, including legal questions about the right of access to a firm's property to monitor, whether prior announcement (and even a warrant) may be required, and how such prior announcement might fit with the discharge adjustment period. Another complication is the development of remote monitoring equipment which may make physical visits inside the plant unnecessary.

Conclusion

It has been argued in this book that environmental economists and policy analysts have been too simplistic in their treatment of monitoring for compliance with discharge standards.[10]

- By ignoring unavoidable variation in the rate of production and the performance of production and treatment processes and the existence of measurement imprecision, they have seriously underestimated the difficulty of determining compliance with a standard and have even ignored the question of how the standard itself ought to be defined.
- By playing down the source's incentive to cheat, they have justified a view that monitoring is not really necessary for the vast majority of sources. Whether this in fact is so remains an open, empirical question.

After the above consideration of the quality control approach to monitoring, however, the reader may be inclined to sympathize with these simplifications. Bringing in source variability and measurement imprecision makes for a difficult problem, one that quality control "solves" only by assuming away some further difficulties.

Indeed, within the constraints set by legal requirements for notice (of inspection and sampling) as described in chapter 3, cumbersome and often batch-test technology, and the reality of discharge variability, it is hard to be sanguine about the real effect of the U.S. system of environmental regulation. On the other hand, this is an area in which technology is developing quite rapidly in a promising direction—toward continuous-reading instruments that can be made to report directly to remote stations, with computer programs designed to do all necessary calculations and comparisons automatically (Ruger, 1981).[11] In addition, there are possibilities for simpler design criteria that are not so demanding of information, and these are described in chapter 7.

---

[10] The chapter has concentrated on standards, but the problems of monitoring for compliance with an effluent charge approach are the same in technical and statistical terms. Thus, another common error in the literature has been to make too much of the distinction between charges and standards in the monitoring dimension. There are differences for overall enforcement purposes, however, as discussed in the chapter 4 literature review.

[11] The existence of a large body of continuously generated data does not mean that statistical questions will vanish, however, since the problem of statistically identifying "violations," perhaps in the presence of autocorrelation, will remain (Gardenier, 1982).

APPENDIX 6-A

Derivation of the Objective Function Expression for Quality Control Sampling Design

Refer back to figure 6-1 in the text. Note that since T_a is the length of time the process is in control and T_r the time the process is out of control then γ, the expected proportion of time the process is out of control, in the long run approaches $E(T_r)/[E(T_a) + E(T_r)]$ where $E(\cdot)$ as usual denotes expectation. Furthermore,

$$E(T_r) = E(T_2) + E(T_3) + E(T_4) \tag{A-1}$$

Time period T_2, the time from the process going out of control to a sample being taken, is by definition:

$$T_2 \equiv s - T_1 \tag{A-2}$$

Time period T_1 is the time to the occurrence of an upset within the interval between samples, s, the mean of which Duncan (1956) shows is:

$$E(T_1) = \frac{1 - (1 + \lambda s)e^{-\lambda s}}{\lambda(1 - e^{-\lambda s})} \tag{A-3}$$

The expected value of T_2 is therefore:

$$E(T_2) = s - E(T_1) = \frac{s}{1 - e^{-\lambda s}} - \frac{1}{\lambda} \tag{A-4}$$

Time period T_3 represents the time elapsed between the first sample taken after occurrence of the upset and the last sample prior to its detection and is obtained from the average out-of-control run length, L_r, of the procedure; that is:

$$E(T_3) \simeq (L_r - 1)s \tag{A-5}$$

The time period T_4 between the last sample prior to detection of lack of control and the end of the cycle is:

$$E(T_4) = D + \tau n \tag{A-6}$$

where:

   $D \equiv$ the average time taken to find and correct an assignable cause for the failure

   $\tau \equiv$ a factor representing the computational delay inherent in sampling. It equals the delay per observation between taking a sample and factoring the sample results into the control scheme. In this formulation, the total delay grows linearly with the number of observations, an overly simple and probably pessimistic assumption

Now all the information needed to write $E(T_r)$ and hence γ in a computationally useful way is in hand. Substituting the above expressions for $E(T_2)$, $E(T_3)$, and $E(T_4)$ into the expression for $E(T_r)$ gives:

$$E(T_r) = \frac{s}{1 - e^{-\lambda s}} - \frac{1}{\lambda} + (L_r - 1)s + D + \tau n \qquad \text{(A-7)}$$

But $E(T_a) = \dfrac{1}{\lambda}$ and $\gamma = \dfrac{E(T_r)}{E(T_r) + E(T_a)}$

so $\gamma \simeq \dfrac{\dfrac{s}{1 - e^{-\lambda s}} - \dfrac{1}{\lambda} + (L_r - 1)s + D + \tau n}{\dfrac{s}{1 - e^{-\lambda s}} + (L_r - 1)s + D + \tau n} \qquad \text{(A-8)}$

This expression shows how γ depends on the parameters of the monitoring scheme, directly through n and s and indirectly as n affects L_r.

Deriving the expression for g is relatively easy. By definition the process goes out of control once per cycle of T_c units of time (hours, days, or whatever metric is chosen). Therefore, g, the average number of occasions the process goes out of control per unit of time, is:[1]

$$g = \frac{1}{E(T_c)} = \frac{1}{\dfrac{s}{1 - e^{-\lambda s}} + (L_r - 1)s + D + \tau n} \qquad \text{(A-9)}$$

[1] The alert reader will note from equations A-8 and A-9 that $\gamma = 1 - g/\lambda$ must be true. At first sight this may appear absurd. Since g is the number of out-of-control (violation) events per unit time and γ is the fraction of each *cycle* spent out of control (in violation), it seems counter intuitive to have the latter depend negatively on the former. This relationship, however, results because the length of the cycle itself changes when g changes. The logic is really best seen in reverse. Taking action to decrease $E(T_r)$ by, for example, increasing test precision and hence decreasing L_r, out-of-control run length, will reduce γ. For a fixed λ this will imply more *incidents* per unit time but less time in violation.

It is almost as easy to work through the logic of the expression for ℓ, the expected number of false alarms per unit time of operation. (This logic is presented clearly in Goel and coauthors, 1968.)

If $E(T_a)$ is the expected time in control and s is the sampling interval, then $E(T_a)/s$ is the expected number of samples taken while the process is in control. But $E(T_a)$ equals $1/\lambda$. Therefore, $1/\lambda s$ represents the expected number of samples taken between initiation of the process and the time it goes out of control. On the average there will be one false alarm after each L_a samples, for L_a is the average run length when the process is in control. Or otherwise stated, the probability of a false alarm *before* the process goes out of control is $1/L_a$. Therefore, the expected number of false alarms per cycle, a_f, is:

$$a_f \equiv \frac{1}{\lambda s} \cdot \frac{1}{L_a} \tag{A-10}$$

The expected number of false alarms per unit time is therefore:

$$\ell = \frac{a_f}{E(T_c)} = \frac{1}{\lambda s} \cdot \frac{1}{L_a} \cdot \frac{1}{E(T_c)}$$
$$= \left[(\lambda s L_a) \left(\frac{s}{1 - e^{-\lambda s}} + (L_r - 1)s + D + \tau n \right) \right]^{-1} \tag{A-11}$$

$$\pounds = \left[\frac{s}{1 - e^{-\lambda s}} - \frac{1}{\lambda} + (L_r(\mu_r,n) - 1)s + D + \tau n \right]$$
$$\times \frac{[B_a(\mu_a) - B_r(\mu_r)] + W + (\lambda s L_a)^{-1} Y}{\left[\frac{s}{1 - e^{-\lambda s}} + (L_r(\mu_r,n) - 1)s + D + \tau n \right]} + \frac{b + cn}{s} \tag{A-12}$$

References

Bagshaw, Michael, and Richard A. Johnson. 1975. "The Effect of Serial Correlation on the Performance of CUSUM Tests II," *Technometrics* vol. 17, no. 1 (February) pp. 73–80.

Bauer, Peter, and Peter Hackl. 1978. "The Use of MOSUMs for Quality Control," *Technometrics* vol. 20, no. 4 (November) pp. 431–436.

———— and ————. 1980. "An Extension of the MOSUM Technique for Quality Control," *Technometrics* vol. 22, no. 1 (February) pp. 1–7.

Berthouex, P. M., and W. G. Hunter. 1975. "Treatment Plant Monitoring

Programs: A Preliminary Analysis," *Journal of the Water Pollution Control Federation* vol. 47, no. 8 (August) pp. 2143–2156.

————, ————, and L. Pallesen. 1978. "Monitoring Sewage Treatment Plants: Some Quality Control Aspects," *Journal of Quality Technology* vol. 10, no. 4 (October) pp. 139–149.

Chiu, W. K., and G. B. Wetherill. 1973. "The Economic Design of Continuous Inspection Procedures: A Review Paper," *International Statistical Review* vol. 41, no. 3, pp. 357–373.

Downing, Paul B., and James Kimball. 1976. "Enforcing Administrative Rules: The Case of Water Pollution Control." Working paper (Blacksburg, Virginia, Virginia Polytechnic Institute and State University, Department of Economics).

Duncan, Acheson, Jr. 1956. "The Economic Design of X-Charts Used to Maintain Current Control of a Process," *Journal of the American Statistical Association* vol. 51 (June) pp. 228–242.

Ewan, W. D. 1963. "When and How to Use CUSUM Charts," *Technometrics* vol. 5, no. 1 (February) pp. 1–22.

Gardenier, Turkan K. 1982. "Moving Averages for Environmental Standards," *Simulation* vol. 39, no. 2 (August) pp. 49–58.

Gibra, Isaac N. 1975. "Recent Developments in Control Chart Techniques," *Journal of Quality Technology* vol. 7, no. 4 (October) pp. 183–192.

Goel, A. L., S. C. Jain, and S. M. Wu. 1968. "An Algorithm for the Determination of the Economic Design of X-Charts Based on Duncan's Model," *Journal of the American Statistical Association* vol. 63 (March) pp. 304–320.

————, and S. M. Wu. 1971. "Determination of A.R.L. and a Contour Nomogram for CUSUM Charts to Control Normal Mean," *Technometrics* vol. 13, no. 2 (May) pp. 221–230.

———— and ————. 1973. "Economically Optimum Design of CUSUM Charts," *Management Science* vol. 19, no. 11 (July) pp. 1271–1281.

Goldsmith, P. L., and H. Whitfield. 1961. "Average Run Lengths in Cumulative Chart Quality Control Schemes, *Technometrics* vol. 3, no. 1 (February) pp. 11–20.

Guttman, Irwin, S. S. Wilks, and J. Stuart Hunter. 1971. *Introductory Engineering Statistics* (New York, John Wiley & Sons), chapters 10 and 11.

Harrington, Winston. 1981. *The Regulatory Approach to Air Quality Management: A Case Study of New Mexico* (Washington, D.C., Resources for the Future).

Kelman, Steven. 1961. *What Price Incentives? Economists and the Environment* (Boston, Auburn House).

Kemp, Kenneth W. 1961. "The Average Run Length of the Cumulative Sum Chart When a V-Mask Is Used," *Journal of the Royal Statistical Society, Series B* vol. 23, no. 1, pp. 149–153.

Kneese, Allen V. 1984. *Measuring the Benefits of Clean Air and Water* (Washington, D.C., Resources for the Future).

Reynolds, Marion R., Jr. 1975. "Approximations to the Average Run Lengths in Cumulative Sum Control Charts," *Technometrics* vol. 17, no. 1 (February) pp. 65–71.

Rice, James K. 1980. "Analytical Issues in Compliance Monitoring," *Environmental Science and Technology* vol. 14 (December) pp. 1455–1457.

Ruger, David W. 1981. "Automatic Reporting of Continuous Monitoring System Data," in *Continuous Emission Monitoring—Design, Operation and Experience* (Pittsburgh, Pennsylvania, Air Pollution Control Association) pp. 253–267.

Vaughan, William J., and Clifford S. Russell. 1983. "Monitoring Point Sources of Pollution: Answers and More Questions from Statistical Quality Control," in *The American Statistician* vol. 37, no. 4, pp. 476–487.

Wetherill, G. Barrie. 1977. *Sampling Inspection and Quality Control*, 2d ed. (London, Chapman and Hall).

7
Lessons from Game Theory Approaches

Earlier chapters have stressed that because of the combination of inevitable fluctuations in pollution discharges and the imprecision of measurement instruments, environmental protection agencies must allow for stochasticity when monitoring. But stochasticity is only a complication overlaying the fundamental issue of compliance, and efforts to design monitoring programs that begin with an assumption of compliance and deal only with stochasticity, such as the model of chapter 6, are as limited in their own way as those that assume away uncertainty over the results of monitoring activities (for example, Storey and McCabe, 1980). Yet bringing these two threads together is difficult; the resulting models yield only grudgingly to analysis. Furthermore, the lessons of the models relevant to the agency's problem—who to monitor and how much— are generally dependent on the assumed availability of information, such as the external damages produced by specific sources, that only an extreme optimist would foresee being available within a decade or two.

In this situation, it is worth asking if further elaboration within the traditional optimization framework is the only route likely to provide insights into the design of agency monitoring policy. The answer seems to be no, and the following sections put forth an alternative approach with a game-theoretic structure. This approach takes into account both the source's option to choose willful noncompliance and the inherent uncertainty of the agency's knowledge about the source's behavior. One of its most promising features is seen to be the independence of rules for agency behavior from measures of the damages caused by the externality. However, this initial single-play game seems to lead to impractically expensive advice to the agency. In particular, the desirable frequency of monitoring is so high, under reasonable assumptions about error probabilities, penalty structures, and the polluter's costs of discharge control, that attaining such a frequency for all sources almost certainly would more than exhaust the agency's monitoring budget, perhaps many times over. To deal with this problem, a game of repeated plays is envisioned. Sources are categorized according to their behavior, as determined by monitoring, and the monitoring frequencies vary with the categories. Under such a scheme it is possible in principle to reduce noncompliance to very low levels even while staying within a very limited monitoring budget.

What is not dealt with here is the possibility of "timed" cheating in which noncompliance begins as soon as monitoring ends and is itself interrupted as soon as a monitoring visit is about to take place. With infrequent, announced monitoring visits, this kind of behavior cannot be ruled out—but neither can it be effectively modeled. To proceed with these rather simple models for designing monitoring schemes it is necessary to assume that control interruptability, or the will to use it, has its limits.[1]

Monitoring as a Single-Play Truth Game

It is not hard to accept the general characterization of the relationship between, say, an environmental protection agency and the polluters it regulates as a game. After all, if the agency promulgates a discharge standard for a source, the decision maker controlling that source at least has the choice of trying to comply or of flouting the standard. On the other side, the agency can choose to monitor the performance of the source or not. Each can be seen as having to choose a strategy in the absence of knowledge of the other's choice, a particular pair of choices producing a particular pair of payoffs, dependent on such features of the problem as the agency's skill at monitoring, its costs, the damages of uncontrolled emissions, the source's control costs, and the penalties levied by the agency for discovered violations.

One route for translating this rather vague but appealing notion into operational terms is provided by an intriguing paper of Brams and Davis (1983) dealing with the problem of verifying behavior in the context of

---

[1] The question of degree of interruptability—the cost and timing implications of control equipment shutdown—is a complicated empirical one not dealt with in this book. It represents an opportunity for further research.

arms control agreements. It will be instructive to consider their model of verification decisions—"the truth game"—because later in the chapter a monitoring game will be built upon its foundations.

There are two parts of the Brams-Davis (BD) truth game: the payoff matrix and the ordering and background of the plays. It is the second part that most strongly suggests the potential link to the monitoring problem, but the first is not without interest. In the BD truth game a signaller (S) announces in some fashion that he is complying with an arms control agreement. His announcement is either true or not true. The other player, the detector (D), chooses whether or not to believe the signaller. The payoff matrix may be written as follows:

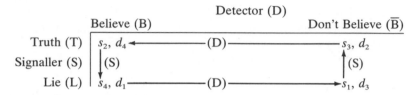

where s_i is the payoff to signaller and d_j the payoff to detector. Each player ranks the payoffs in declining subscript order ($s_4 > s_3 > s_2 > s_1$). Notice, first, that the game is nonzero sum, as the total payoff to the two players must be higher when S tells the truth and is believed than when S lies and is not believed. The labelled arrows indicate another feature of the game: the lack of a stable outcome in the following sense. If S announced a policy of truth telling, D always would choose to believe. But if D announced that he would always believe, S would find lying preferable to truth telling. But then D would find disbelief preferable to belief, shifting S's optimal policy to truth telling . . . and so forth. The same cyclicality exists if s_3 and s_2 or d_3 and d_2 are transposed.

Detailed analysis of this game without the additional BD structure will not be undertaken in this chapter, but it is worth noting that certain probabilistic strategies have interesting properties.[2] In particular, for games having this structure, when actual numerical values are chosen for the payoffs, it is possible to find two probabilities with the following properties:

- If S announces that he will tell the truth with probability t^+ just greater than a number t, depending only on the payoffs available to D, then it will be optimal for D always to believe S, even

---

[2] Brams and Davis (1983, pp. 10–11) provide a few comments and references relevant to the game defined by this matrix.

though S will not always be truthful; that is, when S tells the truth with probability t^+, D's payoff will be highest when he always believes. By always disbelieving or by disbelieving with some probability, D can lower S's payoff but cannot raise his own.

- If D announces that he will believe whatever S says with probability m^-, just less than a number m, depending only on S's payoffs, then it will be optimal for S always to tell the truth. Optimal here has the same meaning as above for t; that is, when D has announced m^-, S can never do better than to tell the truth, although by lying always or sometimes, the payoff of the strategy to D can be reduced.

This idea of a preannounced probabilistic strategy as an inducement to desired behavior by the other party will arise again when the game has been restructured to reflect the monitoring problem.

First, however, consider the second key part of the game, the assumptions about how it is played. The discussion to this point has assumed simultaneous choice of strategies, possibly after preannouncement of probabilistic rules. This is the common notion of a game. But the BD truth game is actually played in a different way. First, S signals compliance. Second, D consults his detection equipment (monitoring instruments) that confirms or contradicts S's assertion. Then D decides whether or not to believe his equipment. And finally, D announces belief or disbelief. This sequential strategy choice with intervening measurement brings us closer to the pollution monitoring situation. Notice further that only when D's detection equipment is imperfect does the game continue to have interest. In particular, if D's equipment can detect the actual state of S's compliance with certainty, then the game has an equilibrium. S can do no better than to tell the truth and be believed. If he lies he will be found out and disbelieved. Thus, the Brams-Davis paper concentrates on the situation in which D's equipment gives the correct answer only with probability $p < 1$, and the truth game looks even more as though it might be useful in the monitoring context.[3] The resulting convergence is made more obvious by a decision-tree repre-

---

[3] Notice that the implication that somehow *both* players can know what actually happened in a play, and therefore what payoffs have actually been "received," is at best problematic. Once D announces belief or disbelief, S knows where he came out, but D, by the nature of the game, cannot know whether he was right or wrong. Only the existence of a deus ex machina revealing the truth about S's actions can resolve D's doubts *unless* S is able and willing to perform this task. This conceptual difficulty remains even after the sequential play structure is introduced. In the monitoring game, however, it is not a problem.

sentation of the full truth game model, in which D, with given probabilities, can accept or reject the readings of his equipment.

Thus, let

t = the probability that S chooses to tell the truth

p = the probability that D's equipment correctly senses the state of S's compliance

q = the probability that D chooses to accept his equipment's reading when it indicates truth telling (compliance) by S

r = the probability that D chooses to accept his equipment's reading when it indicates noncompliance by S.[4]

The decision-tree description of this game is presented in figure 7-1. Notice that D can be right about S either because he believes his equipment when it is right or because he disbelieves it when it is wrong.[5] Just as important, notice that two sorts of errors can occur. D can disbelieve S when S is in fact telling the truth or he can believe S when S is in fact lying. If S's compliance is thought of as constituting a null hypothesis, error I amounts to a false rejection (a false positive); and error II to a false acceptance (a false negative).[6]

As appealing as this picture may be, there are certain features that cannot be accepted if the aim is a model to help inform the design of monitoring systems. Most obviously, the BD truth game builds in the assumption of continuous monitoring in the sense that in every period S chooses and D measures. It is precisely the question of how often D should monitor that is of interest in the pollution context. And for this aspect of the problem to be addressed, it is necessary to look at the costs of monitoring, a notion treated only implicitly by BD. More broadly, if the key decision is whether or not to monitor, the payoff matrix must reflect the costs to each party that logically follow on that decision.

---

[4] Brams and Davis actually define r to be the probability that D chooses to believe S's claim of compliance (disbelieve his equipment) when the reading indicates a violation. The definition in the text will be easier to transfer into the real monitoring game.

[5] D's choice whether or not to believe his equipment may strike the reader as odd. Brams and Davis introduce it for their own purposes, and it plays no important role in the following adaptation of their model. The results depend crucially on the existence of instrument error but not on the possibility of caprice in interpreting instrument readings (See footnote 6).

[6] The error probabilities are defined as conditional on the choice made by S. If D always accepts his equipment's readings, both error probabilities are equal to $1 - p$. If neither q nor r is equal to 1, the probability of a false positive, given compliance, is $p(1 - q) + (1 - p)r$; and the probability of a false negative, given noncompliance, is $(1 - p)q + p(1 - r)$.

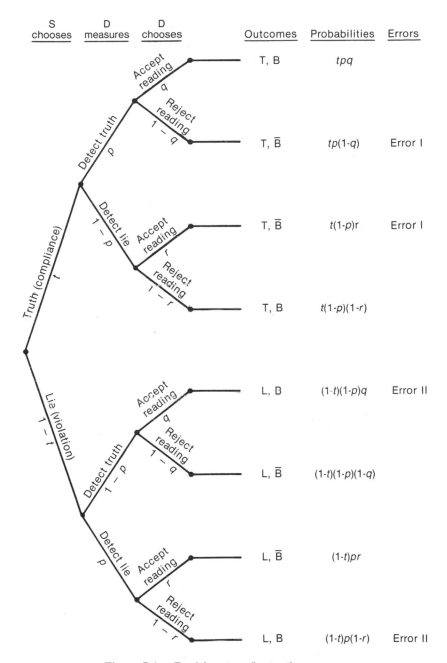

Figure 7-1. Decision tree for truth game

Constructing a Monitoring Game on the
Foundation of the Truth Game

In a monitoring game, the players are the source (S) and the responsible environmental agency (A). The agency will be presumed to have established an upper limit on the source's emissions per period.[7] The source is presumed to have the choice each period of attempting to meet the standard or not. For further simplicity of exposition, assume that the violation choice is taken simply by turning off the control equipment. (This assumption is not central to the most interesting results described but does allow a payoff matrix to be constructed quite easily.) The environmental agency has the choice each period of monitoring or not monitoring the source's discharge. The available monitoring instrumentation is imperfect, in the sense that it produces the correct answer only with probability $p < 1$ when S is in compliance and $z < 1$ when S is in violation (p and z could be equal). The "answer" given by the machine is of the form "violation" or "no violation"; that is, the signal is dichotomous. As in the truth game, the agency can choose to believe or not believe the instrument, and, as above, the probabilities of the agency believing its equipment will be taken to be

q–when the instrument shows compliance, and

r–when the instrument shows a violation.

Again, q and r could be equal. Maintaining a difference merely allows for the possibility of agency bias—which might be thought of as a surrogate for instrument bias.

In a play of the game the source decides whether or not to comply and the agency decides whether or not to monitor. These decisions are made simultaneously, not sequentially. If monitoring occurs and a violation is discovered, the agency fines the source and orders it to start up its control equipment, and the source is assumed both to pay the fine and comply with the order.

The payoffs to the two parties are in the form of costs and may be summarized conveniently as follows (all in per-period terms). For the source:

CC = the cost of controlling discharge to meet the emission standard

---

[7] Such an upper limit could be established as the sum of a desired mean emission rate, μ, plus an allowance for unavoidable emission variation, σ. The upper limit in emissions per period could be $\mu + k\sigma$. Alternatively, the upper limit could be set through legislation without reference to the characteristics of the source. Then source compliance, where discharge variation was unavoidable, would involve the source establishing a target mean emission rate $\mu_T = \mu_L - k\sigma$, where k would be the source's choice (See chapter 6).

X = the extra costs entailed by within-period startup of the control equipment.[8]

F = the fine levied by the agency for a violation discovered

For the agency, as guardian of the public interest:

M = the cost of monitoring emissions

D = the extra damages from uncontrolled as opposed to controlled emissions

The decision-tree description of the game defined by the above assumptions is presented in figure 7-2.

Producing a payoff matrix for the monitoring game requires that for each player the four outcomes produced by the combinations of the two players' choices be valued:[9]

Strategy of S	Strategy of A	$t=$	$m=$	Payoff to S	Payoff to A
Comply	Monitor	1	1	$CC + F(\alpha)$	M
Comply	Don't monitor	1	0	CC	0
Violate	Monitor	0	1	$(CC + F + X)(1 - \beta)$	$M + D(\beta)$
Violate	Don't monitor	0	0	0	D

Here,

$\alpha = tmp(1 - q) + tmr(1 - p) = tm(p - pq + r - rp)$
$\quad$ = probability of type I error

$\beta = (1 - t)m[z(1 - r) + (1 - z)q]$
$\quad$ = probability of type II error

Thus, the payoff (cost) matrix for the monitoring game:

$$
\begin{array}{c c c}
 & \text{Agency (A)} & \\
 & \text{Monitor} & \text{Don't Monitor} \\
\text{Comply} & CC + F(\alpha),\ M \xrightarrow{\quad\quad (A) \quad\quad} & CC,\ 0 \\
\text{Source (S)} & & (S) \downarrow \\
\text{Violate} & (CC + X + F)(1 - \beta),\ M + D(\beta) & 0,\ D
\end{array}
$$

[8] It is assumed in what follows that when a violation is discovered, equipment startup costs plus the full normal costs of control will be incurred during that period. This is expositionally convenient but not crucial, as the payoff magnitudes can always be adjusted by assuming relative sizes for X and F.

[9] See appendix 7-A for derivation. In this and subsequent discussions it is assumed that the source(s) and the agency are risk neutral so that expected monetary values are appropriate decision variables. This seems reasonable in the monitoring and enforcement context when firm survival is not at stake.

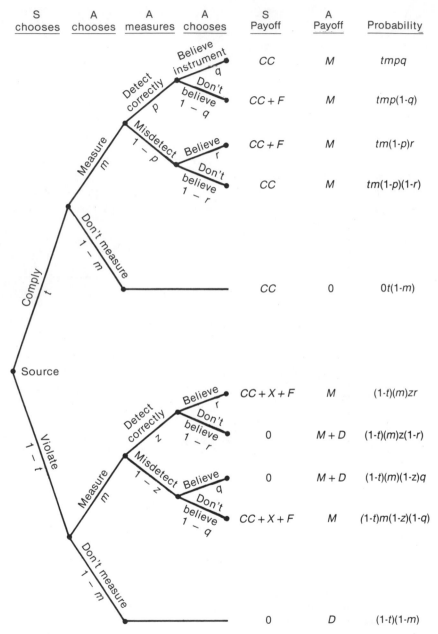

Figure 7-2.   Decision tree for the monitoring game

Notice, first, that cyclicality of the players' preferences over the states is not automatically guaranteed in this game. The two arrows indicate that part of a cycle certainly exists, but it is necessary to ask under what conditions the other two legs exist as well. Under what conditions does A prefer monitoring over not monitoring when S violates? Or, when is $M + D(\beta) < D$? This is true when $M < D(1 - \beta)$, which will be more likely true the smaller M (the cost of monitoring), the larger D (damages from uncontrolled pollution) and the smaller β (the probability of false negative monitoring results)—that is, the higher the "power" of the monitoring test.

The second such question is, when does S choose comply over violate, when A chooses monitor? This will be true when

$$CC + F(\alpha) < (CC + X + F)(1 - \beta)$$

which may be rewritten as

$$CC(\beta) + F(\alpha) < (X + F)(1 - \beta)$$

This amounts to a condition that the expected value of the extra costs of *not* complying (the fine and extra operating costs weighted by the probability of their imposition given that monitoring occurs) exceed the expected extra costs of complying (the basic control costs times the probability that even if monitoring takes place, a violation will not be discovered; plus the fine times the probability of its unjust imposition). Clearly, the higher X and the lower α and β the better the chance of cyclicality. Unless $(1 - \beta) < \alpha$, increasing F will also increase the chance of cyclicality. Since α and β are both commonly expected to be well less than $1/2$, it seems likely that the normal case will be that increasing F increases the chance of cyclical preferences.

Therefore, it is reasonable to assume that both conditions hold and that the monitoring game commonly will have the same structure as the BD truth game.[10] How can this correspondence be exploited? One promising route is to look at the potential for strategic interaction—in particular, to ask whether the agency can announce a strategy that will make compliance optimal for the source. That the answer is yes should not be surprising, given the existence of a truth-inducing mixed strategy

---

[10] If neither cyclicality condition holds, then the source will find violation preferable to compliance no matter what the agency does, and the agency will choose not to monitor.

in the truth game.[11] Such a strategy has obvious intuitive appeal, for it confines S's play to the comply row of the payoff matrix, while reducing the agency's expected costs below the full cost of monitoring in each period. What it lacks is social optimality. There is no reason to expect full compliance to be optimal. But this relaxation of the usual requirement permits a potentially very important relaxation in information requirements, for the quantification of the compliance-inducing mixed strategy will be seen not to depend on D, the incremental damages attributable to noncompliance. Since this is not a piece of data the agency realistically can expect to have, it is of some comfort to know that at least some appealing second-best strategy can be achieved without it.[12]

To derive the compliance-inducing strategy for A the expected value of the game to S is manipulated, reflecting the probabilities t (of compliance) and m (of monitoring). Thus, referring to the payoff matrix for the monitoring game, it is possible to derive the expected value of one play of the game for S. This turns out to be:

$$E(S) = t[CC - (CC + (CC + X + F)(1 - \beta) - (CC + F(\alpha)))m]$$
$$+ (CC + X + F)(1 - \beta)m \tag{1}$$

For $t = 1$ (compliance) to be optimal for S, the expression in brackets must be less than zero. (Recall that S wants to *minimize* expected costs.) That is,

$$[CC - ((CC + X + F)(1 - \beta) - F(\alpha))m] < 0$$

This requirement will be met when:

$$m > \frac{CC}{(CC + X + F)(1 - \beta) - F(\alpha)} \tag{2}$$

Therefore, A must announce a strategy of monitoring with probability m^+, just greater than that defined by the above ratio, which is based entirely on the costs of S and the monitoring capabilities of A.

---

[11] A mixed strategy in this context is one in which the actor sometimes chooses one alternative and sometimes the other.

[12] This is the sense in which the above definition of noncompliance—as uncontrolled discharge—does not matter to the substance of the argument.

It is worth noting that not only does this analysis take as given the cost of monitoring in a period, M, and the error characteristics of the monitoring instrumentation and sampling scheme, α and β, but it ignores the possibility that by changing M the agency might well be able to change α and β. This simplification makes it much easier to set out the logic of the game approach but is clearly inadequate as a representation of the world facing real state agencies. While it is not easy to find empirical evidence supporting the common sense notion that by spending more on better equipment it ought to be possible to reduce α and β, it is at least certain that by spending more on taking and analyzing more samples, with the same equipment, the agency can reduce β for given α (see chapter 5, table 5-2). Thus, a natural question is, how should the agency choose the monitoring method it will employ?

Unfortunately, there is no entirely satisfactory answer available, at least not while maintaining the reduction in information requirements gained by abandoning social optimality for a compliance-inducing mixed strategy. The agency cannot decide among alternative costs and error combinations unless it knows D, the damages from uncontrolled emissions. To see this, examine the expected value of the game to the agency, $E(A)$. This is:

$$E(A) = Mtm + 0[t(1 - m)] + [M + D(\beta)](1 - t)m \qquad (3)$$
$$+ D(1 - t)(1 - m)$$

Rewriting and cancelling where possible, this becomes:

$$E(A) = mM + mD(\beta)(1 - t) - mD(1 - t) + D(1 - t) \qquad (4)$$
$$= mM + D(1 - t)(m\beta - m + 1)$$

If it is assumed that the agency can affect β by its choice of M, it makes sense to ask about the optimal choice of M—where optimal means the value that minimizes expected cost. An appealing specification of the relation between M and β is that

$$\beta = g(M), \quad \text{where } \frac{d\beta}{dM} < 0; \qquad \frac{d^2\beta}{dM^2} > 0$$

so that the relationship is as sketched on the following page.

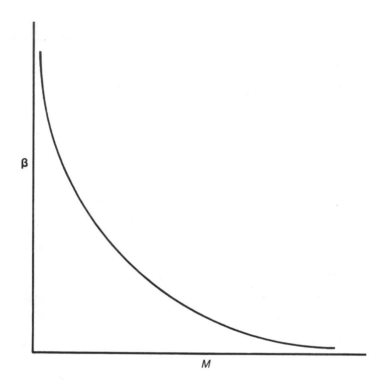

The extreme point of $E(A)$ with respect to M is then the point where:

$$\frac{dE(A)}{dM} = m + D(1 - t)\, m\, \frac{d\beta}{dM} = 0$$

$$\text{or} \quad \frac{d\beta}{dM} = \frac{1}{D(t - 1)} \tag{5}$$

The important point is that finding the extreme point (the optimal method) requires knowing D. But note also that were D known, the agency would choose higher M and lower β, the smaller its estimate of t, the probability of compliance. This much at least is reasonable.[13] The simplified agency problem, however, involves fixed M and β.

Before exploring the likely *minimum* size of m, the compliance-in-

---

[13] For a fuller discussion of the agency's optimal choice problem, with the same underlying difficulty, see chapter 6.

ducing monitoring probability, it is worth asking whether m ever is likely to be greater than one and thus to constitute an absurd requirement.[14] The answer is that as long as the game exhibits the cyclicality noted above, m must be less than one; that is, so long as S prefers comply to violate when A monitors, m will be less than one. Recall that S prefers complying over violating when $CC < (CC + X + F)(1 - \beta) - F(\alpha)$, so that $m < 1$ would be assured.

To explore the likely lower bounds on m, as an indication of the practicality of a compliance-inducing policy in the face of a monitoring budget constraint, it will be convenient to express X and F as multiples of CC. Let $X = kCC$ and $F = \ell CC$. Then:

$$m > \frac{1}{(1 + k + \ell)(1 - \beta) - \ell\alpha} \tag{6}$$

Then if, for example, $k = 0$, $\ell = 1$

$$m > \frac{1}{2(1 - \beta) - \alpha} > \frac{1}{2}$$

If $k = \ell = 1$

$$m > \frac{1}{3(1 - \beta) - \alpha} > \frac{1}{3}$$

But if $k = \ell = .25$

$$m > \frac{1}{\frac{3}{2}(1 - \beta) - \alpha} > \frac{2}{3}$$

And the smaller the error probabilities, the more closely m is approximated by

$$\frac{1}{1 + k + \ell}$$

---

[14] If the expression requires m to be greater than one, the agency cannot induce truthfulness even with constant monitoring. m will never be less than zero as long as $0 < \alpha < 1 - \beta < 1$.

Table 7-1.   Values of the Probability of Monitoring Required to Induce
Compliance

| Case | Value of parameters | | | | Value of m |
	k	ℓ	α	β	
Base	.2	.2	.05	.20	0.901
1	.2	.2	.05	.05	0.758
2	.2	.2	.01	.01	0.722
3	.2	1.0	.05	.20	0.585
4	.2	1.0	.05	.05	0.481
5	.2	1.0	.01	.01	0.460
6	.2	2.0	.01	.01	0.318
7	.2	3.0	.05	.20	0.312
8	.2	3.0	.01	.01	0.242
9	.2	4.0	.05	.20	0.252

Some values of m for combinations of k, ℓ, α, and β are summarized
in table 7-1. (Notice that one version of the currently popular noncompliance penalty involves a fine equal to the cost savings from noncompliance. Under such a policy $\ell = 1$.)

Thus, where fines are sufficiently draconian or where the combination
of potential fine and extra costs of intermittent control operation is
sufficiently great, it is possible for m to be quite low. Values in the
neighborhood 0.25 to 0.30 would seem to be achievable, even with
monitoring equipment displaying substantial imprecision in the face of
the normal variation in discharge rates. Yet even these low values imply
large increases in monitoring activity over current rates and may very
well be higher than legislatures are willing to see funded. For example,
if limits on total daily emissions are in question, for an announced m
of 0.25 to be believable, each source would have to be monitored, on
average over the long run, once every four days. In the current state of
monitoring equipment design, this implies expenditures for, say, air
pollution enforcement alone, of several thousand dollars per *week* per
source. (See chapter 2 and appendix 2-A for estimates of per-visit costs.)
Currently the most common rate of monitoring visits by state agencies
to large air pollution sources is about once per year.

This suggests that rules of thumb based on single-play monitoring
games may not be very helpful, however entertaining their derivation.
One logical route to additional insight does suggest itself, however.
Might not the agency discriminate among sources on the basis of past
performance in the context of a multiple-play game and thus produce
a new source of compliance incentives? Again the answer turns out
to be yes.

A Repeated Monitoring Game with Performance Grouping of Sources

The idea to be pursued in this section is an extension of a scheme suggested by Greenberg (1984) to discourage tax cheating through the use of audits. Greenberg's problem is very similar to the one just identified: that is, while some probability $p_i(y)$ of audit is enough to encourage taxpayer i with income y to report truthfully, there is insufficient money in the tax authority budget to audit at a rate such that no one would have an incentive to cheat. He shows that in the context of a repeated audit game it is possible to devise a scheme such that the fraction of taxpayers cheating can be kept arbitrarily small. The difference between his problem and the one growing out of the monitoring game is that his audits are perfect instruments so that $\alpha = \beta = 0$. It will be shown below that introducing nonzero errors complicates the required scheme but does not appear to vitiate the usefulness of his solution.

That solution rests on a three-part categorization of taxpayers (sources). In group G_1, the probability of an audit is half the value of ρ, itself defined by Greenberg as the lower bound on the set of truth-inducing audit probabilities, $p_i(y)$. Thus, no one in G_1 has the incentive to report truthfully, at least not for a single play of the embedded game. However, those audited in G_1 and found cheating are transferred to group G_2. Here the probability of audit is less than $\rho/2$, as will be explained in a moment. Certainly the incentive in G_2 *in the single-play game* is not to comply. But if a taxpayer in G_2 is audited and found to be cheating, that person is transferred to group G_3 in which audits are certain and from which there is no escape. If a taxpayer in G_2 is audited and found in compliance that taxpayer is returned to G_1.

The categories and paths between them are shown schematically in figure 7-3. The solid lines refer to the errorless version of the game.

For convenience in his proofs, Greenberg assumes that each taxpayer faces infinitely many (undiscounted) future plays of the tax-audit game. Therefore, absorbtion into G_3 is effectively a sentence to perpetual tax truthfulness. The chance ever to be in G_1, where cheating (and hence a chance of private gain) is optimal, requires the taxpayer to avoid G_3 with certainty; that is, to comply when in G_2 even though the audit probability may be far lower than required to make compliance optimal in the single-play version of the game. (Hence the lack of a path from G_2 to G_3 in the errorless game.) Thus, the key to compliance in G_2 is not current audit frequency but the threat of an infinite (undiscounted)

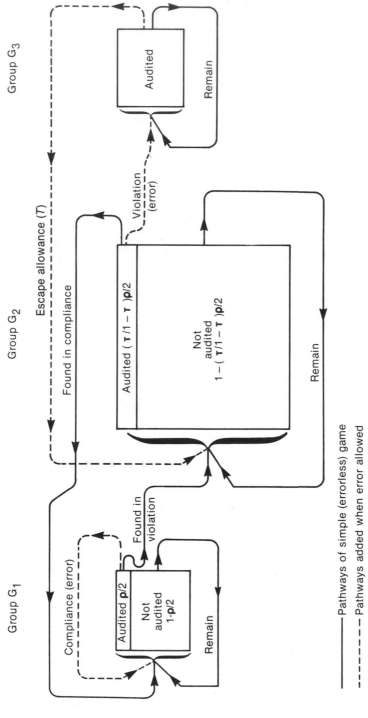

Figure 7-3. Schematic of multiple-play monitoring game (with and without error)

future of complete auditing.[15]

The other feature of Greenberg's model is the assignment of fractions of the population to the groups. Let τ be the fraction assigned initially to G_1 and $1 - \tau$ the fraction assigned initially to G_2. (Then the fraction assigned initially to G_3 is zero.) τ is defined as follows:

$$0 < \tau < \text{Min} \left[\frac{1}{2}, \varepsilon, \frac{r}{\rho} \right] \tag{7}$$

where r is the audit probability achievable for the group as a whole in view of the budget constraint, ρ is as defined above, and ε is the target fraction of cheaters. Under this assignment rule the fraction of cheaters, those in G_1, initially will be less than the target ε. In addition, it is easy to show that the fraction of the population audited initially will be less than r, the constraint set by the budget.[16] Furthermore, if every taxpayer perfectly follows his self-interest, cheating in G_1 and reporting truthfully in G_2, then G_3 remains always empty, and the probability of audit and hence of movement between groups define a Markov process with two states, and a transition matrix as follows:

	G_1	G_2
G_1	$1 - \rho/2$	$\rho/2$
G_2	$\left(\dfrac{\tau}{1 - \tau}\right)\dfrac{\rho}{2}$	$1 - \left(\dfrac{\tau}{1 - \tau}\right)\dfrac{\rho}{2}$

---

[15] Greenberg shows that with discounting but still an infinite time horizon, the fraction of taxpayers cheating can be kept below a level ε, which depends on the size of the discount rate. Or, said another way, if ε is the target fraction cheating, discount rates ($\delta = 1/1 + i$) cannot be above $\delta(\varepsilon)$.

[16] The fraction of the total population audited is

$$\tau \frac{\rho}{2} + (1 - \tau)\left(\frac{\tau}{1 - \tau}\right)\frac{\rho}{2} = \rho\tau$$

If 1/2 is the minimum of the three terms in the assignment function, then

$\rho\tau < \rho/2$ and $1/2 < r/\rho$ so $\rho\tau < \rho/2 < r$

If ε is the minimum, then $\rho\tau < \rho\varepsilon$ and $\varepsilon < r/\rho$ so $\rho\tau < \rho\varepsilon < r$

And finally, if r/ρ is the minimum, $\rho\tau < \rho(r/\rho) < r$ is true.

The stationary probability vector for G_1 and G_2 (the fraction of the population in the two groups after the game has been repeated many times) is found from the equations (for example, Parzen, 1960, pp. 136 and the following):

$$\Pi_1 = (1 - \rho/2)\Pi_1 + \left(\frac{\tau}{1 - \tau}\right)(\rho/2)\Pi_2$$

$$\Pi_2 = (\rho/2)\Pi_1 + \left[1 - \left(\frac{\tau}{1 - \tau}\right)\frac{\rho}{2}\right]\Pi_2$$

$$\Pi_1 + \Pi_2 = 1 \tag{8}$$

to which the solution is[17]

$$\Pi_1 = \tau; \quad \Pi_2 = 1 - \tau$$

so that the original allocation remains stable over the long run.[18]

---

[17] Greenberg's specification of starting proportions and his design of audit probabilities are thus simply a convenient way of insuring that the requirements on frequency of cheating on the budget constraint are met from the beginning.

[18] Greenberg's model can lead to some peculiar results over at least some range of values of p_2, the audit probability in G_2. Because everybody complies in G_2 to avoid transition to perpetual audit in G_3, increasing the probability of audits in G_2 only increases the transition rate to G_1, where everybody cheats, everything else equal. Thus, if the audit probability in G_1 is p_1 and G_2 is p_2, then the transition matrix for a single play of the game is

	G_1	G_2
G_1	$(1 - p_1)$	p_1
G_2	p_2	$(1 - p_2)$

for which the stationary vector is

$$\Pi_1 = \frac{p_2}{p_1 + p_2} \quad \text{and} \quad \Pi_2 = \frac{p_1}{p_1 + p_2}$$

and Π_1 is the long-term fraction of cheaters.

But $d\Pi_1/dp_2 = \dfrac{p_1}{(p_1 + p_2)^2} > 0$

Then, for any $\hat{p}_1$, there will be an upper limit to $\hat{p}_2$ as determined by the budget constraint. But there will be no lower limit short of zero. And as p_2 declines, both overall audit costs and the proportion of cheaters in the population decline. This peculiar result occurs

Because pollution monitoring instruments are imperfect, however, even perfect pursuit of self-interest in the infinitely repeated game will not prevent some sources in G_2 from being identified as cheaters when monitored. (The version of the game with error taken into account is displayed in figure 7-3 using dashed lines for the paths.) Thus, the membership of G_3 in general will not be zero after any finite number of plays of the game. Indeed, if G_3 is left as an absorbing state as in the Greenberg scheme, audit errors eventually will result in the entire population being in it. Thus, the scheme would fail with an imperfect audit instrument because eventually the budget constraint would be violated as sources accumulated in G_3, each of which had to be audited with probability one. Since otherwise the idea of categorization and the use of a multiple-period future threat seems promising, it is worth exploring the possibility of introducing an escape from G_3 designed to balance in the long run the "recruitment" resulting from auditing errors.

As a first step in this exploration, notice how the existence of false negatives and positives alters the transition matrix for the multiperiod game. Keeping the same rule for initial allocation of sources to G_1 and

---

because the taxpayers (or polluters) are really looking only at the infinite stream of future audits. Any nonzero probability of receiving such a sentence will produce compliance.

A slightly different approach in this same vein is that proposed by Landsberger and Meilijson (1982). They use a two-state system in which the audit probabilities are different in the two states. Failing an audit in the state with the lower probability results in a transfer to the state with a higher probability. The major concern of these authors is to show that their system is preferable both to a single-state system having the single-audit frequency equal to the weighted average of the p_1 and p_2 for states one and two and to a static two-state system with p_1 and p_2 audit probabilities but no transition between states. The measure of performance involves the tax collection agency's revenue experience, not the compliance rate in the yes/no sense. But in the process they provide some additional insight into the case of interest here.

In particular, they work out in some detail the case for constant income across all taxpayers and no discounting of future net income. In this situation there is a lower bound on p_2. It must be high enough to induce truth telling in order that state two not be absorbing with everybody belonging to it, everybody cheating, and the budget constraint violated nonetheless. But it is also possible for p_2 to be too high, so that excessively frequent audits do nothing to improve compliance. On the contrary, the excessive frequency simply pushes a greater proportion of the population back into G_1, where they can cheat. Thus, only over a range of values of p_1, and hence of p_2, will the two-state system be preferred to one with a single-state and an audit rate lower than that required to produce compliance. Thus, in a sense, the effect of adding the third state or group is to move down the range of values of p_2 for which the multistate system has an advantage. But in both cases, the *best* p_2 is the lowest one in the feasible range.

G_2 (no sources initially assigned to G_3) and the same audit probabilities—$\rho/2$ in G_1 and $\tau\rho/2(1 - \tau)$ in G_2—the transition matrix becomes:

	G_1	G_2	G_3
G_1	$1 - \dfrac{\rho}{2}(1 - \beta)$	$\dfrac{\rho}{2}(1 - \beta)$	0
G_2	$\dfrac{\rho\tau(1 - \alpha)}{2(1 - \tau)}$	$1 - \dfrac{\rho\tau}{2(1 - \tau)}$	$\dfrac{\rho\tau(\alpha)}{2(1 - \tau)}$
G_3	0	?	?

As before, β is the probability of a false negative when a violation is occurring, and α is the probability of a false positive when compliance is occurring. Because the probability of transferring from G_2 to G_3 is no longer zero, it is no longer possible simply to leave out the last row and column of the matrix as was done in Greenberg's tax-avoidance model. But how to structure that row is the question. One way that certainly will not work is to make G_3 absorbing:

	G_1	G_2	G_3
G_3	0	0	1

The question marks in the full matrix indicate that allowing for some transition back to G_2 may allow use of a modified Greenberg scheme to encourage compliance even with imperfect monitoring. The requirements that must be met in replacing those question marks are:

- The Markov process should be stationary, with the proportions of sources in G_1, G_2, and G_3 in the long run independent of how the process is started.
- The budget constraint represented by the maximum overall audit frequency, r, must be met in the long run.
- The maximum fraction of sources that can be expected to be in violation must be ε in the long run.

Furthermore, since the sources in G_3 all will be monitored with probability 1 and all thus will be complying (if acting in a rational, self-interested way), the escape probability must be arbitrary; it cannot be tied to success in the audit, or almost all sources would escape each period ("almost" because of false positives). Expected residence time in G_3 would be slightly more than one period, and it would hardly constitute a sufficient penalty to inspire compliance in G_2, with its low probability of monitoring. It is possible to imagine many possibilities tied to *past* behavior, not to behavior in G_3, but these are not analyzed here. They remain matters for further investigation.

To see if it is possible to find a transition probability out of G_3 into G_2 (call it T) such that the requirements are met, begin by writing the transition matrix in simpler terms. Define:

$$Q = \frac{\rho}{2}(1 - \beta)$$

$$P = \frac{\rho}{2}\left(\frac{\tau}{1 - \tau}\right)\alpha$$

so that the matrix becomes:

	G_1	G_2	G_3
G_1	$1 - Q$	Q	0
G_2	$\left(\dfrac{1 - \alpha}{\alpha}\right)P$	$\dfrac{\alpha - P}{\alpha}$	P
G_3	0	T	$1 - T$

Using the equations defining the stationary probability vector, Π_i, the requirement that the overall audit probability be no greater than r, and substituting r/ρ for τ, the initial fraction of sources assigned to G_1, it is possible to solve for T. That escape probability will guarantee the fulfillment of the budget constraint. Notice, however, that by this substitution for τ, $\varepsilon > r/\rho$ is implicitly assumed. What this amounts to is an implicit agreement to accept Π_1, the long-run fraction of the sources in violation, as sufficiently small. This is because an ε is not specified. Rather, it is assumed that the desired limit on the fraction of cheaters is large enough to be attained with given r/ρ.

The algebra for this solution is found in appendix 7-A. Here the answer is simply written down and its interpretation discussed. The solution says that for

$$T < \frac{(1 - \beta)\alpha\rho(r - 1)}{(1 - \alpha)(\rho - 2r) - (1 - \beta)(\rho - 2r)} \tag{9}$$

it is possible for the scheme to work in the sense just defined. But notice, first, that if $\alpha = \beta$, the upper limit on T is not defined; and, second, that it is possible for the upper limit to be negative, a nonsense result, since T is a probability. To guarantee that the upper limit is meaningful, because $r - 1 < 0$ must be true, it is necessary for the denominator to be negative as well. This will be true when either:

$$(1 - \beta) > (1 - \alpha) \text{ [both are } > 0] \text{ and } \rho > 2r \qquad (10)$$

or

$$(1 - \alpha) > (1 - \beta) \text{ and } \rho < 2r \qquad (10a)$$

To see what these conditions mean, observe from the appendix that the ratio:

$$\frac{\Pi_3}{\Pi_1} = \frac{\frac{\rho}{2}(1 - \beta)\alpha}{(1 - \alpha)T} \qquad (11)$$

Thus, other things being equal, when $(1 - \beta) > (1 - \alpha)$, the fraction of sources in G_3, where the audit frequency is 1, independent of ρ, is higher. In that circumstance, ρ can be higher than when the opposite holds; that is, when $(1 - \alpha) > (1 - \beta)$, so that the proportion of sources with audit frequencies dependent on ρ is relatively greater.

Notice that, so long as the parameter values produce a positive upper limit for T, where $t < 1$ is required, it does not matter what T is actually chosen. In particular, as smaller values of T are chosen, Π_3 increases, Π_1 decreases, and the available funds are more completely exhausted. Thus, let:

$$\alpha = 0.05, \quad \beta = 0.20, \quad \rho = 0.312, \quad r = 0.20$$

so that $(1 - \alpha) > (1 - \beta)$ and $\rho < 2r$. Then from expression (9) $T < 0.753$ must be true. But if $T = 0.05$

$$\Pi_1 = 0.115$$
$$\Pi_2 = 0.870$$
$$\Pi_3 = 0.015$$

and the overall audit frequency is:

$$\frac{\rho}{2} \Pi_1 + \frac{\rho}{2} \left(\frac{\tau}{1-\tau}\right) \Pi_2 + \Pi_3 = 0.048 < r$$

While if $T = 0.01$:

$\Pi_1 = 0.109$

$\Pi_2 = 0.820$

$\Pi_3 = 0.070$

and $\dfrac{\rho}{2} \Pi_1 + \dfrac{\rho}{2} \left(\dfrac{\tau}{1-\tau}\right) \Pi_2 + \Pi_3 = 0.101 < r$

Thus *given* the assumption that the prospect of a long (though not permanent) stay in G_3 is sufficient incentive for sources to comply in G_2, it has been shown that it is possible to devise a workable rule for releasing some arbitrarily chosen sources from G_3. The rule is workable in that it allows satisfaction of the budget constraint when the stationary probability vector defines the probabilities of being in each of the states. An obvious question, however, is whether the prospect of a limited (expected) period of certain auditing is likely to be an effective incentive.

An answer to this question can be obtained by going back to a single play of the monitoring game with the new information from the multiple-play game in hand. In particular the key question is how the source would behave when in G_2. For that play, the source can choose to comply or not, and it may be audited or not. If it is audited and complying, it faces a large probability $(1 - \alpha)$ of going to G_1 and suffering no penalty and a small probability (α) of being sent to G_3 with an expected prospect of a long stay under complete monitoring. If it is not audited, it stays in G_2 and faces the same choice again. If it chooses not to comply, again it may escape without audit or may be the beneficiary of a falsely negative audit. But it may be audited and sent to G_3. If compliance provides the lower expected costs at one play it should do so at every play in G_2.

For simplicity, but not with any idea that the result is definitive, the following explores this notion in the absence of discounting. Thus, figure 7-4 displays the appropriate decision tree, in which the returns from the single play itself are distinguished from the expected future returns that would ineluctably accrue to the source if that branch were travelled. (The key probability, of leaving G_3, is again denoted as T.) The tran-

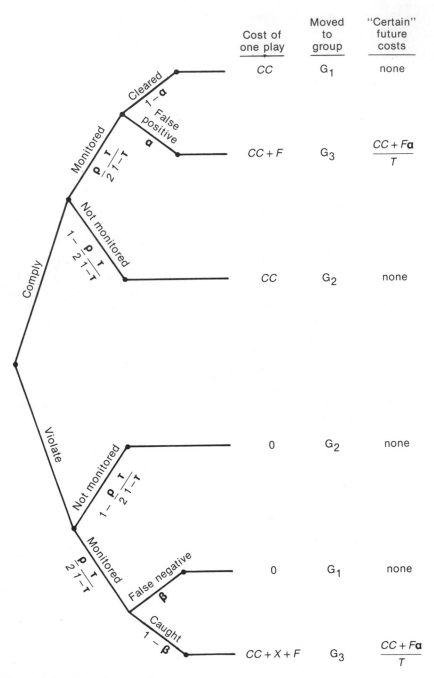

Figure 7-4. Decision tree for the monitoring game when a source is in G_2

sitions are also noted. From this diagram it is possible to construct the expected value to the source of the two strategies, comply and violate.

$$E(C) = \frac{\rho}{2}\left(\frac{\tau}{1-\tau}\right)\left[(1-\alpha)CC + \alpha(CC + F) + \alpha\left(\frac{CC + F\alpha}{T}\right)\right]$$

$$+ \left(1 - \frac{\rho}{2}\left(\frac{\tau}{1-\tau}\right)\right)(CC) \tag{12}$$

$$E(V) = \frac{\rho}{2}\left(\frac{\tau}{1-\tau}\right)(1-\beta)\left[CC + X + F + \frac{CC + F\alpha}{T}\right] \tag{13}$$

It can be shown (and is in appendix 7-A) that for $E(C) < E(V)$ and compliance to be preferred to violation, the following must hold:

$$T < \frac{(1 - \alpha - \beta)(1 + \ell\alpha)\rho\tau}{2 - 3\tau} \tag{14}$$

It is of some comfort to note that this requirement is satisfied by the second example displayed just above, where

$$\alpha = 0.05, \quad \beta = 0.20, \quad T = 0.01, \quad \ell = 3, \quad \tau = 0.1, \quad \rho = .312$$

Then $T = 0.01 < \dfrac{(0.75)(1.15)(0.0312)}{2 - 0.3} = \dfrac{.0269}{1.7} = 0.016$

Conclusions

Thus there is at least a preliminary case for internal consistency. A value for T can be found such that the Markov process is stationary, the budget constraint is satisfied, and, at least without discounting, the incentive for occupants of G_2 is to comply. This all suggests that improving the monitoring scheme design may not be completely hopeless, even when budget constraints prevent monitoring that would be frequent enough to give all sources the incentive to comply at all times. Suitable categorization based on observed behavior can reduce the fraction of non-complying sources to very low levels while still meeting the budget constraint.

The result also has the advantage of suggesting numerous routes to be explored in the future with an eye to widening understanding of the circumstances in which such a scheme can work. In particular, a fuller exploration of the conditions under which internal consistency can be maintained will give a better feeling for the practical applicability of the idea. A further obvious extension is to a system of emission charges as opposed to upper limit emission standards. Other explorations of the differential implications of monitoring and enforcement arrangements under the two sorts of implementation policies have found intriguing differences (see, for example, Linder and McBride, 1984). The situation is unlikely to be different in the context of multiple-play games, although just how important such differences are as a practical matter, with charges playing essentially no part in actual policy, could be questioned.

One other line of inquiry has been suggested by Brams, who wrote that "there is a more general issue in your work and ours—the use of threats to achieve certain ends" (personal correspondence, 25 June 1984). In Brams (1984), Brams and Hessel (1984), and Brams and Merrill (1984), the use of threats in sequential games (single-play but multiple-sequential choices of strategy allowed) is analyzed. In these games, the threatener is willing to settle for a lower outcome himself in order to punish the person threatened. Basically, the threat says: give me what I want or I will force you to choose between two outcomes low on your list of ordinally ranked rewards. This is not possible in every game setup, of course.

The questions this raises are these. Is it possible to characterize the multiple-play game as a threat game? Can this be done without making the situation excessively artificial? Will this process improve the intuitive appeal of the game device?

Finally, it should be stressed that the framework developed in this chapter is potentially applicable to a range of public monitoring problems where cost, instrument error, and opportunities for noncompliance combine to make a difficult resource allocation problem—and where traditional assumptions that source-specific damage functions are known simply cannot be taken seriously.

APPENDIX 7-A

Derivation of Expressions Used in Text

1. Expected payoffs in the single-play game.

For Comply/Monitor

$$
\begin{aligned}
\text{Payoff to S} \quad &= CC(pq) + (CC + F)p(1 - q) + (CC + F)(1 - p)r \\
&\quad + CC(1 - p)(1 - r) \\
&= CC(pq + p - pq + r - pr + 1 - p - r + pr) \\
&\quad + F(p - pq + r - pr) \\
&= CC + F[p(1 - q) + (1 - p)r] \\
&= CC + F(\text{probability of a false positive}) \\
&= CC + F(\alpha) \quad\quad\quad\quad\quad\quad\quad\quad\quad\quad\quad\text{(A-1)}
\end{aligned}
$$

$$
\begin{aligned}
\text{Payoff to A} \quad &= M(pq) + M(p(1 - q)) + M(1 - p)r \\
&\quad + M(1 - p)(1 - r) \\
&= M \quad\quad\quad\quad\quad\quad\quad\quad\quad\quad\quad\quad\quad\quad\quad\text{(A-2)}
\end{aligned}
$$

Notice that the same results are obtained here and in what follows if $r = q = 1$ so that $\alpha = 1 - p$ and $\beta = 1 - z$.

For Violate/Monitor

$$
\begin{aligned}
\text{Payoff to S} \quad &= (CC + X + F)(zr) + 0[z(1 - r)] + 0(1 - z)q \\
&\quad + (CC + X + F)(1 - z)(1 - q) \\
&= (CC + X + F)(zr + 1 - z - q + zq) \\
&= (CC + X + F)[1 - z(1 - r) - (1 - z)q] \\
&= (C + X + F)(1 - \text{probability of false negative}) \\
&= (CC + X + F)(1 - \beta) \quad\quad\quad\quad\quad\quad\quad\quad\text{(A-3)}
\end{aligned}
$$

$$
\begin{aligned}
\text{Payoff to A} \quad &= M(zr) + (M + D)[z(1 - r)] + (M + D)(1 - z)q \\
&\quad + M(1 - z)(1 - q) \\
&= M(zr + z - zr + q - zq + 1 - z - q + zq) \\
&\quad + D[z(1 - r) + (1 - z)q] \\
&= M + D(\beta) \quad\quad\quad\quad\quad\quad\quad\quad\quad\quad\quad\quad\text{(A-4)}
\end{aligned}
$$

For Comply/Don't Monitor

$$
\text{Payoff to S} = CC \quad\quad\quad\quad\quad\quad\quad\quad\quad\quad\quad\quad\quad\quad\text{(A-5)}
$$

$$
\text{Payoff to A} = 0 \quad\quad\quad\quad\quad\quad\quad\quad\quad\quad\quad\quad\quad\quad\quad\text{(A-6)}
$$

For Violate/Don't Monitor

Payoff to S $= 0$ (A-7)

Payoff to A $= D$ (A-8)

2. Expected value of the game to the source.

$$
\begin{aligned}
E(S) &= [CC + F(\alpha)]tm + (CC + X + F)(1 - \beta)(1 - t)m \\
&\quad + CC[t(1 - m)] + 0(1 - t)(1 - m) \\
&= [CC + F(\alpha)]tm - CCtm - (CC + X + F)(1 - \beta)tm \\
&\quad + CCt + (CC + X + F)(1 - \beta)m \\
&= t[CC - (CC + (CC + X + F)(1 - \beta) - (CC + F(\alpha)))m] \\
&\quad + (CC + X + F)(1 - \beta)m \\
&= t[CC - ((CC + X + F)(1 - \beta) - F(\alpha))m] \\
&\quad + (CC + X + F)(1 - \beta)m
\end{aligned}
$$ (A-9)

3. Finding the stationary probabilities in the multiple-play game. The transition matrix is:

	G_1	G_2	G_3
G_1	$1 - Q$	Q	0
G_2	$\dfrac{1 - \alpha}{\alpha}P$	$\dfrac{\alpha - P}{\alpha}$	P
G_3	0	T	$1 - T$

The solution for Π_i is obtained from the equations

$$
\Pi_1 = \Pi_1(1 - Q) + \Pi_2\left(\frac{1 - \alpha}{\alpha}\right)P
$$ (A-10)

$$
\Pi_2 = \Pi_1 Q + \Pi_2\left(\frac{\alpha - P}{\alpha}\right) + \Pi_3 T
$$ (A-11)

$$
\Pi_3 = \Pi_2 P + \Pi_3(1 - T)
$$ (A-12)

$$
\Pi_1 + \Pi_2 + \Pi_3 = 1
$$ (A-13)

We have from (A-12), $P\Pi_2 = T\Pi_3$. From expression (A-11) and this relation it is possible to derive the following:

$$\Pi_2 = \Pi_1 Q + \Pi_2 \left(\frac{\alpha - P}{\alpha}\right) + P\Pi_2$$

or

$$\Pi_1 Q = \Pi_2 \left(\frac{\alpha - \alpha + P - \alpha P}{\alpha}\right)$$

so that $\Pi_1 = \dfrac{P}{Q}\left(\dfrac{1 - \alpha}{\alpha}\right)\Pi_2$

Then using expression (A-13),

$$\Pi_2 \left[\frac{P}{Q}\left(\frac{1 - \alpha}{\alpha}\right)\right] + \Pi_2 + \frac{P}{T}\Pi_2 = 1$$

$$\Pi_2 \left[\frac{TP(1 - \alpha) + \alpha QT + \alpha QP}{\alpha QT}\right] = 1$$

and so,

$$\Pi_1 = \frac{(1 - \alpha)PT}{(1 - \alpha)PT + \alpha QT + \alpha QP}$$

$$\Pi_2 = \frac{\alpha QT}{(1 - \alpha)PT + \alpha QT + \alpha QP}$$

$$\Pi_3 = \frac{\alpha QP}{(1 - \alpha)PT + \alpha QT + \alpha QP}$$

Substituting the underlying expressions for Q and P gives

$$\Pi_1 = \frac{(1 - \alpha)\tau T}{(1 - \alpha)\tau T + (1 - \beta)(1 - \tau)T + \frac{\rho}{2}(1 - \beta)\alpha\tau} \tag{A-14}$$

$$\Pi_2 = \frac{(1 - \beta)(1 - \tau)T}{(1 - \alpha)\tau T + (1 - \beta)(1 - \tau)T + \frac{\rho}{2}(1 - \beta)\alpha\tau} \tag{A-15}$$

$$\Pi_3 = \frac{\rho/2(1 - \beta)\alpha\tau}{(1 - \alpha)\tau T + (1 - \beta)(1 - \tau)T + \frac{\rho}{2}(1 - \beta)\alpha\tau} \tag{A-16}$$

To find the expression for U, the overall proportion of the sources audited, begin by writing

$$U = \Pi_1 \frac{\rho}{2} + \Pi_2 \frac{\rho}{2} \frac{\tau}{1 - \tau} + \Pi_3$$

Substituting from expressions (A-14), (A-15), and (A-16) the following is obtained:

$$U = \frac{(1 - \alpha)T\tau \, \rho/2 + (1 - \beta)T\tau \, \rho/2 + (1 - \beta)\alpha\tau \, \rho/2}{(1 - \alpha)T\tau + (1 - \beta)(1 - \tau)T + (1 - \beta)\alpha\tau \, \rho/2} \tag{A-17}$$

and the question of interest is the relation of U to r, the budget limitation.

4. Meeting the budget requirement.

Substituting $\dfrac{r}{\rho}$ for τ and cancelling gives:

$$U = \frac{(1 - \alpha)rT + (1 - \beta)rT + (1 - \beta)\alpha r}{2(1 - \alpha)\frac{r}{\rho}T + (1 - \beta)\left(1 - \frac{r}{\rho}\right)T + (1 - \beta)r\alpha} \overset{?}{\gtreqless} r$$

$$\tag{A-18}$$

Or, clearing and combining terms:

$$U = \frac{(1 - \alpha)\rho rT + (1 - \beta)(\rho r)(T + \alpha)}{(1 - \alpha)2rT + (1 - \beta)[\alpha\rho r + 2T(\rho - r)]} \overset{?}{\gtreqless} r$$

Clearing the major fraction and combining terms in T on the left hand side, this becomes:

$$(1 - \alpha)\rho rT + (1 - \beta)\rho rT - (1 - \alpha)2r^2T - 2(1 - \beta)r(\rho - r)T$$

$$\overset{?}{\gtreqless} - \rho r(1 - \beta)\alpha + (1 - \beta)r(\rho\alpha r)$$

Thus, after cancelling r,

$$T \overset{?}{\gtreqless} \frac{(1 - \beta)(r - 1)\rho\alpha}{(1 - \alpha)(\rho - 2r) - (1 - \beta)(\rho - 2r)} \tag{A-19}$$

But the numerator of this solution for T is negative, because $r < 1$. Therefore, if U is to be less than r *and* the upper limit on T is to be positive, it must be true that:

$$(1 - \alpha)(\rho - 2r) - (1 - \beta)(\rho - 2r) < 0$$

Thus, if $\rho > 2r$, $(1 - \beta)$ must be $> (1 - \alpha)$ so that the second term dominates; while if $\rho < 2r$, $(1 - \alpha) > (1 - \beta)$ must be true so that the first term dominates.

5. Providing incentive to comply in G_2.
   Beginning with expressions (12) and (13) in the text, let

$$\frac{\rho}{2}\left(\frac{\tau}{1 - \tau}\right) = Z$$

Then $E(C) = Z\left[CC + F\alpha + \alpha\left(\frac{CC + F\alpha}{T}\right)\right]$

$$+ (1 - Z)CC \tag{A-20}$$

and $E(V) = Z(1 - \beta)\left[CC + X + F + \frac{CC + F\alpha}{T}\right] \tag{A-21}$

For $E(C) < E(V)$ to be true, it is necessary that:

$$Z[F\alpha - (1 - \beta)(CC + X + F)$$
$$+ \frac{1}{T}[(CC + F\alpha)(\alpha + \beta - 1)]] < -CC \tag{A-22}$$

Or, substituting, as in the first section of the chapter, ℓCC for F and kCC for X and changing the sign of the inequality, expression (A-22) is transformed into (A-23):

$$Z\left[(1 - \beta)(1 + k + \ell) - \ell\alpha + \frac{1}{T}(1 - \alpha - \beta)(1 + \ell\alpha)\right] > 1 \tag{A-23}$$

But, from the original game, recall that

$$m \; (\geq\!\rho) = \frac{1}{(1 - \beta)(1 + k + \ell) - \ell\alpha}$$

Thus expression (A-23) can be written as:

$$Z\left[\frac{1}{\rho} + \frac{1}{T}(1 - \alpha - \beta)(1 + \ell\alpha)\right] > 1 \qquad\qquad (A\text{-}24)$$

To solve for T such that this will be true, rearrange expression (A-24) as:

$$Z[T + \rho(1 - \alpha - \beta)(1 + \ell\alpha)] > \rho T \qquad\qquad (A\text{-}25)$$

or

$$ZT - \rho T > -\rho(1 - \alpha - \beta)(1 + \ell\alpha)Z$$

which is more understandably written as

$$T < \frac{\rho(1 - \alpha - \beta)(1 + \ell\alpha)Z}{(\rho - Z)} \qquad\qquad (A\text{-}26)$$

But $Z = \dfrac{\rho}{2}\dfrac{\tau}{1 - \tau}$ so it must be true that:

$$
\begin{aligned}
T &< \frac{\rho(1 - \alpha - \beta)(1 + \rho\alpha)\dfrac{\rho}{2}\dfrac{\tau}{1 - \tau}}{\rho - \dfrac{\rho}{2}\dfrac{\tau}{1 - \tau}} \\[2mm]
&= \frac{(1 - \alpha - \beta)(1 + \ell\alpha)\rho\tau}{2 - 3\tau}
\end{aligned}
\qquad\qquad (A\text{-}27)
$$

References

Brams, Stephen J. 1984. "Deterrence and Uncertainty: A Game-Theoretic Analysis," Economic Research Report 84-06. (C.V. Starr Center for Applied Economics, New York University, Department of Economics, New York, March).

———, and Morton D. Davis. 1983. "The Verification Problem in Arms Control: A Game-Theoretic Analysis," Economic Research Report 83-12. (C.V. Starr Center for Applied Economics, New York University, Department of Economics, New York, June).

———, and Marek P. Hessel. 1984. "Threat Power in Sequential Games," *International Studies Quarterly* vol. 28, pp. 23–44.

———, and Samuel Merrill III. 1984. "Binding vs. Final-offer Arbitration: A Combination Is Best." Paper presented at the American Association for the Advancement of Science meetings, May, New York City.

Greenberg, Joseph. 1984. "Avoiding Tax Avoidance: A (Repeated) Game-Theoretic Approach," *Journal of Economic Theory* vol. 32, no. 1 (February) pp. 1–13.

Heyde, C. C. 1983. "Law of the Iterated Logarithm," in Samuel Kotz and Normal L. Johnson, eds., *Encyclopedia of Statistical Sciences* (New York, John Wiley & Sons).

Landsberger, Michael, and Isaac Meilijson. 1982. "Incentive Generating State Dependent Penalty System," *Journal of Public Economics* vol. 19, pp. 333–352.

Linder, Stephen H., and Mark E. McBride. 1984. "Enforcement Costs and Regulatory Reform: The Agency and Firm Response," *Journal of Environmental Economics and Management* vol. 11, no. 4 (December) pp. 327–346.

Parzen, Emanuel. 1960. *Modern Probability Theory and Its Application* (New York: John Wiley & Sons).

Storey, D. J., and P. J. McCabe. 1980. "The Criminal Waste Discharger," *Scottish Journal of Political Economy* vol. 27, no. 1 (February) pp. 30–40.

8

Conclusions and Recommendations

The tone of this book has tended to be critical. The frequency of monitoring visits by responsible agencies to check on continuing compliance by pollution sources has been characterized as inadequate to reveal much about actual patterns of discharge. And the nature of those visits, especially that they commonly are announced in advance, was taken further to vitiate their usefulness. One reason for both the infrequency and preannounced character of the visits was seen to be the complexity and unwieldy character of the batch-sampling measurement methods designed for initial compliance testing. Another was the uncertain state of the law as it applies to right of entry to perform monitoring functions. Furthermore, the review of the economic and policy analysis literatures was implicitly critical of the meager attention given to the problems of monitoring and enforcement. Even the courts came in for some criticism, both for their unwillingness to come to grips with the issue of statistical errors in identifying violations and compliance and for the tendency of some courts to accept an argument that a polluter has a right to privacy in his polluting activities.

But the purpose of the book is not merely to be critical. It is to advance and change the terms of public discussions of monitoring and enforcement policies, as they apply to pollution control legislation and regulations in particular. To do that and to set the record a bit straighter than it may seem at this point, this final chapter begins with a recognition that the Environmental Protection Agency (EPA), and others active in environmental policy, are pursuing improvements in continuing com-

218

pliance; that is, attention *is* shifting from technology installation and initial testing to day-to-day performance.

Some future chronicle of environmental policy may want to set out the relationship between the dramatic events of Reagan's first term, including the resignation under fire of EPA's administrator, and the subsequent path of enforcement activity. From this vantage point it appears that during the Burford years environmentalists became intensely frustrated by the apparent lack of interest in compliance. One outcome was a campaign of citizen law suits to force compliance with the Clean Water Act. This campaign was spearheaded by the Natural Resources Defense Council and the Sierra Club (*Inside EPA*, 1984a). It took advantage of the wealth of self-monitoring data produced by industrial sources, data showing noncompliance by many sources. Some of the violations were not considered significant under EPA rules, but according to *Inside EPA* about two-thirds of the 200 or so suits filed involved significant noncompliance.[1]

In May 1984 EPA expressed concern about the potential embarrassment to be caused by this flurry of suits but at the same time announced its intention to make it possible for citizen law suits to augment its own activities (*Inside EPA*, 1984a). The efficacy of this approach to spurring continuing compliance was enhanced in November 1984 when a U.S. district court rejected a number of common defenses that had been tried by the targets of these suits. The court found that the plaintiffs (a New Jersey and a national environmental group) had standing to sue and that the self-monitoring data used as the basis for the suit had to be taken to be sufficiently accurate to allow the suit to be heard. The court also rejected other technical legal defenses (*Inside EPA*, 1984b).

This external heat then interacted with internal pressures for more attention to continuing compliance. The resulting change in agency view can hardly be put better than it has been by EPA's Wasserman (1984) in the concluding chapter of her report to the Organization for Economic Cooperation and Development on U.S. environmental compliance monitoring and enforcement. Therefore, her comments are quoted extensively here (Wasserman, 1984, chapter VII, pp. 1–3).

> The traditional approach to compliance monitoring worked fairly well when compliance focused on initial installation of pollution control equip-

[1] Significant noncompliance according to draft EPA documents reported on in *Inside EPA* (1984a) is defined in terms similar to those mentioned in chapters 2 and 3 above. That is to say, a percentage allowance is added to the actual permit limitations to define monthly noncompliance as significant or not. Furthermore, if excessive emissions not meeting the monthly "upper upper" limit are chronic events, occurring four out of a period of six months, they become significant noncompliance events.

ment at a limited number of large facilities. Detection was comparatively straightforward. . . .

* * * * *

The traditional approach has run into several problems [among them]:

1. *In the older established programs such as air and water, the problems have shifted from initial compliance to continuing compliance.* While it is impossible to remain in compliance 100% of the time, there are few norms to indicate what reasonable expectations should be. Moreover, periodic inspections are no longer sufficient to assure that plants are maintaining sound operations in conformance with environmental requirements over an extended period of time. This places greater emphasis on source self-monitoring and reporting, and simple hand held instruments that support more frequent inspections at problem sources. . . .

* * * * *

Several improvements are needed in the existing compliance monitoring and enforcement systems in the U.S. Some have begun to be made. These include:

• Simple means of identifying and describing requirements for compliance;
• Improved compliance monitoring methods development;
• Improved inspection targeting techniques;
• Additional and more streamlined authority to impose administrative penalties; and simpler techniques for calculating penalties that will withstand challenge . . .

This book has attempted to contribute to the redesign and improvement called for by Wasserman through discussion of the interrelated legal, technical, statistical, and economic aspects of the overall problem. In each of this book's chapters, specific conclusions have been spelled out, and it is not the purpose of this final chapter to summarize or repeat this material; rather, it will draw together a few major themes. Accordingly, the next section addresses themes and makes recommendations relevant to monitoring and enforcement policy, while the second takes a similar line with respect to research.

Policy

A major conclusion of this study, as of Wasserman's paper, must be that the emphasis in law and regulation on achieving the installation of desirable treatment technologies (initial compliance) has been overdone.

The effects of this concentration have extended from the design of monitoring instruments, through the operational definition of the terms compliance and noncompliance, to the measurement of enforcement activity. A correspondingly pervasive change in emphasis and attitude is necessary if, in the coming decades, the nation is to get its money's worth from its substantial investment in pollution control hardware. The heart of this change must be a shift to concentration on *continuing compliance*—that is, the day-to-day, week-to-week, and month-to-month compliance of sources with their permit terms.[2]

This broad change implies in turn several more specific requirements. The most obvious is that attention must shift from the measurement of *ability* to meet permit terms in a special one-time test situation to the frequent measurement of *actual performance*. This is necessary whether or not sources are required to perform self-monitoring, because only by actual measurement can the accuracy and, indeed, truthfulness of self-monitoring reports be checked. Successful achievement of this shift will require changes in technology and clarification, if not actual changes, in law.

The instrumentation for monitoring must be made less cumbersome and less demanding of skilled labor inputs. This will reduce both the effective notice given to a source about to be monitored and the cost of a monitoring visit (or, said another way, the cost of a particular number of samples of a waste stream). One might think of this as an effort to reduce cost per sample substantially enough that even if the standard error of each sample measurement rises significantly, the precision of the measurements resulting from a visit at least will not be worse than now, while the cost of the visit will fall substantially.

As was observed in chapter 3, one path to this goal, the use of remote monitoring equipment, is currently subject to uncertainty because of confusion over whether remote measurements might represent an unreasonable invasion of privacy. This confusion is only part of a larger one, however, in which the federal courts are trying to find common ground on the question of whether a discharger of pollution should have some expectation of privacy in respect to his discharge activity. It seems fair to label this a confusion because the notion that a creator of a public harm should be able to vary the amount and kind of that harm without *effective* public oversight seems so much at odds with society's behavior in analogous situations. To take only one example, the drunken driver or the driver of an unsound automobile is, in the expectational sense,

---

[2] It is important to emphasize that the introduction of marketability for permits (as in the public and offset policies) does not change anything said here, although it may make the monitoring and enforcement task slightly more difficult.

a creator of a public harm. Neither character is protected from the prying eyes of the police in their efforts to prevent the actual occurrence of such an event. Thus, cars may be stopped and subjected to safety checks and drivers subjected to breathalyzer tests without warrant. The person who takes out a license and drives on public roads implicitly accepts this as part of normal public oversight. Similarly, it seems that the courts ultimately will have to find that the discharger of pollution implicitly accepts oversight in the form of remote or surprise monitoring of his discharges. Once this principle is established it will be possible to talk about monitoring frequency and, particularly, about random monitoring visit frequencies in a meaningful way. So long as substantial advanced notice is required, because of a combination of unclear laws and un-suitable equipment, monitoring visits must be thought of as producing measurements of what the source wants measured.

A second major need for clarification as a basis for a new commitment to enforcing continuing compliance is for the standards to be defined with a uniform, or at very least an explicit, statistical interpretation. This may well imply that the very meaning of compliance and noncom-pliance must be thought through afresh. In the process, all involved, including especially the federal courts, must be persuaded to take se-riously the unavoidable problem of statistical errors—both of false neg-atives and of false positives: there must be general recognition that there is some probability both that a source making no effort to comply may be found in compliance as the result of a random monitoring visit and that a source making a serious effort to comply may be found in violation. Neither probability can be driven to zero in a real situation; and reducing one, leaving everything else the same, increases the other. Both *can* be reduced, or one reduced for constant level of the other, by spending more money on more samples or more precise instruments. Chapter 6 described a model and technique for choosing a sampling scheme to minimize overall costs including those of false violation reports, when the costs of sampling, excess emissions, and fruitless search for the cause of an incorrectly identified violation can be assumed known. In real problems only the first costs are known, and so more or less arbitrary rules of thumb must be used to design sampling schemes. But the ac-ceptance of the principle is the most important matter in any event.

The case summaries in chapter 3 suggest that the route to this ac-ceptance must surmount two obstacles. One is by now familiar to an-alysts in nearly every area of public policy, but especially to those dealing with environmental risks. It is the reluctance of individuals to think in probabilistic terms, and the peculiar rules of thumb they use when forced to do so (see, for example, Fischhoff and coauthors, 1981). Layered over this reluctance, in the case of enforcement, appears to be a feeling that the statistical errors somehow represent unfairness; that is, it seems *unfair* that a source trying to comply can be found in violation, even 1 percent of the time.[3] But such a view leads to a dead end, for knowledge of the source's performance is limited by what can be observed and measured. It is possible that a discharge measurement showing violation could be weighed against an inspection visit outcome showing evidence of proper installation and operation of the proper control equipment. But the source's testimony about its intent can have no weight in the identification of violations. If the enforcing agency had to prove intent to violate in order to levy a penalty, it would be placed in a nearly impossible position. Probabilities of error inevitably accompany reliance on objective measurements. This is simply one outcropping of the deeper problem of human knowledge, not some special difficulty devised to bedevil sources of pollution.

Once the inevitability of errors is accepted it will be easier to deal with the appropriate definition and enforcement of standards. Here the major requirement is to end the use of several levels of ad hoc "factors" that, taken together, produce decision rules with unknown error characteristics. In particular, assume the standard in a permit is to be set on the basis of mean (expected) performance plus some number of standard deviations as estimated from the discharge variations of model plants employing the designated technology (see chapter 3). Then it is unnecessary and unwise to multiply that standard by some other factor greater than one in setting a violation detection limit. This is because the measured discharges from the model plant reflect both underlying discharge variation *and* presumably measurement error of the same sort as will be faced in day-to-day monitoring for compliance. Thus, the ad hoc factors described in chapter 3 just reduce the probability of a false positive and increase the probability of a false negative by unknown amounts. The characteristics of the violation test are then unknown for any particular residual and almost certainly differ across residuals even within the same broad group.[4]

[3] In hypothesis testing it is presumably, then, "unfair" that a true hypothesis is rejected some fraction of the time. But of course people do not have feelings for hypotheses, nor do hypotheses or their inventors normally pay fines when rejection occurs. If they did, the idea of fairness might confuse scientific work, as it now does legal reasoning.

[4] The groups were: I, inorganic and oxygen-demanding pollutants (such as BOD, suspended solids, and nutrients); and, II, toxic pollutants (such as heavy metals, cyanide, and organics).

Research Needs

This section is aimed particularly at economic and policy research, but it must begin with emphasis of something implicit in the policy discussion. Research in monitoring instruments must be encouraged to head in the direction of easily transportable, quickly set-up equipment. The analytical methods at the heart of the instrument should be chosen to require smaller rather than larger inputs of skilled labor. To the extent possible, automated methods involving minimal intervention by site-visit personnel should be chosen. In addition, work on in-situ continuous monitoring equipment should continue with an eye on the ideal, if almost certainly unattainable, tamperproof, continuous, remote-recording instrumentation. To make this commitment a real one, it will be necessary for EPA to abandon the idea that any acceptable monitoring method must compare favorably in precision with the reference methods spelled out in the *Federal Register*. It is not so important what the precision of a method is as that the precision is known and constant in use.

For the design of monitoring and enforcement policies, there are several important areas of research suggested by the evidence and analysis presented in chapters 2 through 7 above. One empirical question, central to an understanding of what actually is possible in enforcement activity and what methods are successful, is: How does continuing compliance vary between states that use a voluntary compliance (no penalty) approach and those that levy significant penalties for detected violations? It may not be at all easy to find the answer, for the reason developed in chapter 1. Knowledge of compliance experience generally is poor exactly because monitoring for continuing compliance is spotty at best. Moreover, the use of self-monitoring records to answer this question involves a logical flaw, since what is reported to the state agency can be presumed to reflect not only actual effort and measurement but also the sources' knowledge of the penalties for reported violations.

A second empirical question, this one central to an assessment of what is achievable, is: What would be the costs and times involved in efforts by sources to approximate the battlement pattern of compliance? Or, how quickly and easily can a source turn on and off its control equipment, and how much money is saved by turning it off? This question involves a comprehensive engineering look at the major control technologies, along with careful economic thought about the identification of real cost savings and extra costs.

An entirely different line of research is suggested by the models, in particular the game-theory-based model of chapter 7. Some examples of questions deserving additional attention before that model can be taken entirely seriously as a basis for policy formulation include:

- How sensitive is the size of the efficiency loss in the single-play context implied by the encouragement of 100 percent compliance as opposed to the socially optimal compliance rate? This can be addressed only by the use of hypothetical models, since the necessary damage functions relating to discharges do not exist. A similar, though more complicated, question may be asked in the multiple-play context.
- How, if at all, is it possible to restructure the multiple-play arrangements to avoid the result that higher audit frequencies for groups G_1 and G_2 result in higher levels of noncompliance? (Recall that this followed because an audit is a not-quite-certain ticket out of G_2, where compliance is optimal even though the audit frequency is not high enough to make it so in a single play of the game, back to G_1, where violation is optimal.)
- What are the alternatives for choosing the escape probability from group G_3 (where auditing occurs every period with certainty)? In the model presented in chapter 7, this probability is entirely arbitrary and constant across sources. Are there ways to make use of past behavior to tailor escape probabilities to individual sources while still retaining the incentive effects that are so appealing in the current structure?
- Over what ranges of the key parameters do the incentive and budget consistency results hold? How sensitive are the results to discounting?
- Are there other structures—more groups, for example, or fine increases instead of monitoring frequency increases—that will achieve the desired results at least as effectively?

It is clear even from this short list of research opportunities that much remains to be done. Again, this book was intended to be only the first step in a series of many.

References

Fischhoff, Baruch, Sarah Lichtenstein, Paul Slovic, Stephen L. Derby, and Ralph L. Keeney. 1981. *Acceptable Risk* (Cambridge, Cambridge University Press).

Inside EPA. 1984a. "Ruckelshaus Worried Citizen Suits Will Reveal Poor Enforcement Record," vol. 5, no. 19 (May 11) pp. 1, 6, 7.
———. 1984b. "District Court Grants Judgement in Citizens Suit on NPDES Violation," vol. 5, no. 48 (November 30) p. 5.
Wasserman, Cheryl. 1984. "Improving the Efficiency and Effectiveness of Compliance Monitoring and Enforcement of Environmental Policies. United States: A National Review." Draft (Organization for Economic Cooperation and Development, Environmental Directorate).

Index

227